ANNALS *of* THE NEW YORK ACADEMY OF SCIENCES

T0310252

ME
.86

ISBN-10: 1-57331-883-3; **ISBN-13:** 978-1-57331-883-9

ISSUE

The Year in Ecology and Conservation Biology

ISSUE EDITORS

William H. Schlesinger and Richard S. Ostfeld

Cary Institute of Ecosystem Studies

TABLE OF CONTENTS

Annals of the New York Academy of Sciences (ISSN: 0077-8923 [print]; ISSN: 1749-6632 [online]) is published 30 times a year on behalf of the New York Academy of Sciences by Wiley Subscription Services, Inc., a Wiley Company, 111 River Street, Hoboken, NJ 07030-5774.

Mailing: *Annals of the New York Academy of Sciences* is mailed standard rate.

Postmaster: Send all address changes to ANNALS OF THE NEW YORK ACADEMY OF SCIENCES, Journal Customer Services, John Wiley & Sons Inc., 350 Main Street, Malden, MA 02148-5020.

Disclaimer: The publisher, the New York Academy of Sciences, and the editors cannot be held responsible for errors or any consequences arising from the use of information contained in this publication; the views and opinions expressed do not necessarily reflect those of the publisher, the New York Academy of Sciences, and editors, neither does the publication of advertisements constitute any endorsement by the publisher, the New York Academy of Sciences and editors of the products advertised.

Publisher: *Annals of the New York Academy of Sciences* is published by Wiley Periodicals, Inc., Commerce Place, 350 Main Street, Malden, MA 02148; Telephone: 781 388 8200; Fax: 781 388 8210.

Journal Customer Services: For ordering information, claims, and any inquiry concerning your subscription, please go to www.wileycustomerhelp.com/ask or contact your nearest office. *Americas:* Email: cs-journals@wiley.com; Tel:+1 781 388 8598 or 1 800 835 6770 (Toll free in the USA & Canada). *Europe, Middle East, Asia:* Email: cs-journals@wiley. com; Tel: +44 (0) 1865 778315. *Asia Pacific:* Email: cs-journals@wiley.com; Tel: +65 6511 8000. *Japan:* For Japanese speaking support, Email: cs-japan@wiley.com; Tel: +65 6511 8010 or Tel (toll-free): 005 316 50 480. Visit our Online Customer Get-Help available in 6 languages at www.wileycustomerhelp.com.

Information for Subscribers: *Annals of the New York Academy of Sciences* is published in 30 volumes per year. Subscription prices for 2013 are: Print & Online: US$6,053 (US), US$6,589 (Rest of World), €4,269 (Europe), £3,364 (UK). Prices are exclusive of tax. Australian GST, Canadian GST, and European VAT will be applied at the appropriate rates. For more information on current tax rates, please go to www.wileyonlinelibrary.com/tax-vat. The price includes online access to the current and all online back files to January 1, 2009, where available. For other pricing options, including access information and terms and conditions, please visit www.wileyonlinelibrary.com/access.

Delivery Terms and Legal Title: Where the subscription price includes print volumes and delivery is to the recipient's address, delivery terms are Delivered at Place (DAP); the recipient is responsible for paying any import duty or taxes. Title to all volumes transfers FOB our shipping point, freight prepaid. We will endeavour to fulfill claims for missing or damaged copies within six months of publication, within our reasonable discretion and subject to availability.

Back issues: Recent single volumes are available to institutions at the current single volume price from cs-journals@wiley.com. Earlier volumes may be obtained from Periodicals Service Company, 11 Main Street, Germantown, NY 12526, USA. Tel: +1 518 537 4700, Fax: +1 518 537 5899, Email: psc@periodicals.com. For submission instructions, subscription, and all other information visit: www.wileyonlinelibrary.com/journal/nyas.

Production Editors: Kelly McSweeney and Allie Struzik (email: nyas@wiley.com).

Commercial Reprints: Dan Nicholas (email: dnicholas@wiley.com).

Membership information: Members may order copies of *Annals* volumes directly from the Academy by visiting www. nyas.org/annals, emailing customerservice@nyas.org, faxing +1 212 298 3650, or calling 1 800 843 6927 (toll free in the USA), or +1 212 298 8640. For more information on becoming a member of the New York Academy of Sciences, please visit www.nyas.org/membership. Claims and inquiries on member orders should be directed to the Academy at email: membership@nyas.org or Tel: 1 800 843 6927 (toll free in the USA) or +1 212 298 8640.

Printed in the USA by The Sheridan Group.

View *Annals* online at www.wileyonlinelibrary.com/journal/nyas.

Abstracting and Indexing Services: *Annals of the New York Academy of Sciences* is indexed by MEDLINE, Science Citation Index, and SCOPUS. For a complete list of A&I services, please visit the journal homepage at www. wileyonlinelibrary.com/journal/nyas.

Access to *Annals* is available free online within institutions in the developing world through the AGORA initiative with the FAO, the HINARI initiative with the WHO, and the OARE initiative with UNEP. For information, visit www. aginternetwork.org, www.healthinternetwork.org, www.oarescience.org.

Annals of the New York Academy of Sciences accepts articles for Open Access publication. Please visit http://olabout.wiley.com/WileyCDA/Section/id-406241.html for further information about OnlineOpen.

Wiley's Corporate Citizenship initiative seeks to address the environmental, social, economic, and ethical challenges faced in our business and which are important to our diverse stakeholder groups. Since launching the initiative, we have focused on sharing our content with those in need, enhancing community philanthropy, reducing our carbon impact, creating global guidelines and best practices for paper use, establishing a vendor code of ethics, and engaging our colleagues and other stakeholders in our efforts. Follow our progress at www.wiley.com/go/citizenship.

ANNALS *of* THE NEW YORK ACADEMY OF SCIENCES

VOLUME
1286

ISSUE

The Year in Ecology and Conservation Biology

ISSUE EDITORS

William H. Schlesinger and Richard S. Ostfeld

Cary Institute of Ecosystem Studies

Published by Blackwell Publishing
On behalf of the New York Academy of Sciences

Boston, Massachusetts
2013

Ann. N.Y. Acad. Sci. ISSN 0077-8923

ANNALS OF THE NEW YORK ACADEMY OF SCIENCES

Issue: *The Year in Ecology and Conservation Biology*

Risks to biodiversity from hydraulic fracturing for natural gas in the Marcellus and Utica shales

Erik Kiviat

Hudsonia, Annandale, New York

Address for correspondence: Erik Kiviat, Hudsonia, P.O. Box 5000, Annandale, NY 12504. kiviat@bard.edu

High-volume horizontal hydraulic fracturing (HVHHF) for mining natural gas from the Marcellus and Utica shales is widespread in Pennsylvania and potentially throughout approximately 280,000 km² of the Appalachian Basin. Physical and chemical impacts of HVHHF include pollution by toxic synthetic chemicals, salt, and radionuclides, landscape fragmentation by wellpads, pipelines, and roads, alteration of stream and wetland hydrology, and increased truck traffic. Despite concerns about human health, there has been little study of the impacts on habitats and biota. Taxa and guilds potentially sensitive to HVHHF impacts include freshwater organisms (e.g., brook trout, freshwater mussels), fragmentation-sensitive biota (e.g., forest-interior breeding birds, forest orchids), and species with restricted geographic ranges (e.g., Wehrle's salamander, tongue-tied minnow). Impacts are potentially serious due to the rapid development of HVHHF over a large region.

Keywords: Appalachian Basin; biodiversity; forest fragmentation; hydraulic fracturing; salinization; shale gas

Introduction

High-volume horizontal hydraulic fracturing (HVHHF) occurs at increasing density across potentially 280,000 km² of the eastern United States underlain at depth by the natural gas–bearing Marcellus and Utica shales. These industrial installations and their edge effects alter as much as 80% of local landscapes.[1] The predicted intensity, speed, and extent of industrialization of the landscape have engendered concern about human health but little discussion of the effects on biodiversity,[2–4] although HVHHF has been identified as a global conservation issue.[5] Although the biota of the eastern United States is relatively well studied, many of the rare organisms potentially susceptible to industrial impacts are not. For example, the woodland salamanders (*Plethodon*) are diverse and sensitive to landscape and soil conditions; many species have only been described in recent decades; and as a group they are declining.[6–8] Although a direct survey of many taxa may be infeasible, indicator taxa may not effectively represent overall diversity.[9] In general, various taxa use different micro- and macrohabitats and have different conservation needs; one taxon may not predict the occurrence or sensitivity to impacts of another taxon.[10] This review focuses on the physical and chemical impacts of HVHHF on habitats, taxa, and guilds, and suggests which organisms have particular sensitivities that may put them at risk.

The Marcellus–Utica region

Conservatively, 9.5% of the conterminous United States is underlain by gas shales;[11] Canada, southern South America, Europe, South Africa, North Africa, China, India, and Australia also have exploitable formations.[12] The most extensive resources in the eastern United States are the Marcellus and Utica shales, underlying the Appalachian Basin from approximately the Mohawk and Hudson rivers in New York, through extensive areas of Pennsylvania and Ohio, most of West Virginia, and into small parts of Maryland, Virginia, and Ontario (Fig. 1).[13] Much of the region is forested, with dominant trees that include oaks (*Quercus* spp.), hickories (*Carya* spp.), sugar maple (*Acer saccharum*), American beech (*Fagus grandifolia*), and yellow birch (*Betula allegheniensis*).[14] Elevations range

doi: 10.1111/nyas.12146

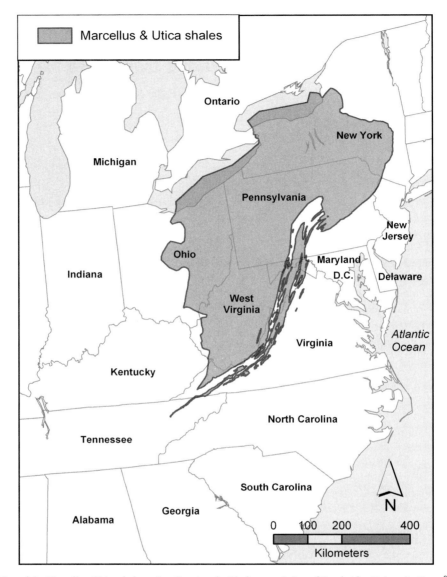

Figure 1. Map of the Marcellus–Utica shale region. Reprinted with the permission of Cambridge University Press.[8]

from less than 100 m near the Hudson River to more than 1500 m in north-central Pennsylvania.

The Marcellus and Utica shales are organic-rich, marine shales deposited during the Middle Devonian and Middle Ordovician periods, respectively. The formations vary from exposed (in small areas) to overlain by 3 km or more of other bedrock strata, with the Utica underlying the Marcellus and extending farther west and southwest. Some of the organic matter is methane, the principal constituent of natural gas, tightly bound in microscopic pores.

Hydraulic fracturing

Horizontal drilling and hydraulic fracturing were developed in recent decades to mine gas from deep strata. In a typical installation, one to several wells are drilled from a single wellpad. Each well descends vertically 1.5 km or more to the target shale stratum, and then continues horizontally as much as 1.5 km. Fracturing fluid (water, chemicals, and sand) is forced under high pressure into the shale to open and prop spaces that let gas flow into the well.[13]

After fracturing, the gas and a portion of the fracturing fluid ascend the well and are collected. The gas is cleaned, compressed, and piped via collector lines to transmission pipelines.

Each HVHHF installation constitutes a wellpad, an access road, storage areas for water, chemicals, sand, and wastewater, a compressor station, and a collector pipeline. Installations often require extensive cut-and-fill, and some are on steep slopes.[15] In Pennsylvania in 2008, half of the installations were in forests and used, on average, 3.56 ha, thereby affecting approximately 15 ha of forest per installation.[1] An estimated 60,000 new wells will be in place by 2030.[16] A well is fractured at intervals of several years during its projected 40- to 50-year life, and each wellpad may support several wells. Each fracturing episode, per well, uses 4–12×10^6 L of water, which is usually trucked from a lake or river (the amount per episode may be as high as 15–25×10^6 L).[17] The portion of water and chemicals that returns to the surface as wastewater has been estimated at 9–100%.[18] More than 600 synthetic chemicals are used in HVHHF, including methanol, napthalene, xylene, acetic acid, ammonia, and #2 fuel oil,[2] but those used in any given well are unidentified. These chemicals constitute about 0.5% of the fracturing fluid; because of the large volume of fluids, 1×10^6 L of chemicals may be injected with a portion returning to the surface.[4,13] The wastewater, either return water during the fracturing operation or produced water afterward,[4] also contains substances from the shale, especially sodium, chloride, bromide, arsenic, barium, other heavy metals, organic compounds, and radionuclides.[13] Wastewater is often stored in lined, open ponds near wellpads, apparently to concentrate it, then trucked to treatment plants (including municipal plants not designed to remove salinity or radionuclides, and discharging effluent that has sometimes led to high salinity or total dissolved solids in rivers).[13,18] Wastewater is also reused for fracturing, disposed of by deep injection, spread on roads for dust control, or concentrated by evaporation and buried.[2,15,18]

Assessing biodiversity risk

Water and soil pollutants
Many spills or leaks of raw chemicals, fracturing fluids, or wastewater have been documented, involving volatile and gaseous organic chemicals, diesel fuel, surfactants, metals, sodium chloride, acidic water, and other substances.[2,3,19–21] In one instance, the median chloride content of wastewater was 56,900 mg L^{-1}.[18] At a West Virginia site, wastewater with approximately 4,000–14,000 mg L^{-1} chloride was sprayed on ground and vegetation, killing trees and other plants.[15] Four northeastern amphibian species have been shown to be adversely affected by approximately 50–1,000 mg L^{-1} chloride, depending on the species and life stage,[22] suggesting that small amounts of HVHHF wastewater could render breeding habitats unsuitable. Many lichens,[23–25] liverworts,[26] sphagnum mosses,[27–29] conifers,[30,31] aquatic plants,[32,33] and bog plants[34] are also sensitive to salt; numerous streams are already salinized from road deicing.[35] Furthermore, lichens[36–40] and stoneworts[41–43] can be harmed by heavy metals. Wastewater ponds contain highly toxic synthetic chemicals[2] and could potentially be ecological traps for water birds, muskrat, turtles, frogs, and aquatic insects. Mixtures of these chemicals will have effects that cannot be predicted by knowledge of individual chemicals.[3]

Sediment pollution of streams and other habitats may be caused by heavy equipment on rural roads mobilizing mineral particles in runoff or airborne dust,[13] or by inadequate erosion control at HVHHF sites.[21] In an HVHHF region of Arkansas, stream turbidity was correlated with well density.[3] Suspended sediment additions to higher order streams could potentially harm benthic invertebrates and fish; native brook trout and freshwater mussels are among the most vulnerable taxa. Dust from roads can harm nearby plants and pollute streams.[35]

Forest loss and fragmentation
Loss of forest cover and change in the spatial pattern of cover are often confounded, but cause different responses.[44] Edge effects on forest biota range from 10 m for trees to as much as 500 m for certain birds.[45] Forest fragmentation, which affects dispersal, pollination, herbivory, and predation, is a major conservation concern in HVHHF landscapes;[1,16,46] 20% or more of the forest cover may be removed for the establishment of HVHHF installations, and more than 80% of the land may be affected if a 100-m edge effect is considered.[1] This loss and fragmentation of forest would result in the warming and drying of the remaining forest, with greater penetration by nonnative plants, songbird nest predators, and the brood-parasitic

brown-headed cowbird (*Molothrus ater*). Several forest amphibians occur at lower abundances in forest within 25–35 m of clearcut edges,[47] and juvenile forest amphibians have trouble dispersing across open habitats.[48,49] At five conventional gas well sites in West Virginia, three salamander species were more abundant closer to the forest edge, but less so in the drier southwestern aspect than in the moister northeastern aspect; edge effect was offset by rock and coarse woody debris (CWD) microhabitats.[50] Organisms sensitive to forest fragmentation include lichens and bryophytes,[51] orchids,[52] other herbs,[53] the West Virginia white butterfly (*Pieris virginiensis*),[54] amphibians,[8,48,55] and birds.[56–59] Orchids are among the taxa most sensitive to habitat change in that many orchid species occur in small, isolated populations and depend on narrow ranges of soil moisture, organic matter, light, and nutrients; they also have complex obligate relationships with mycorrhizal fungi and pollinators.[60] In addition, drying of air and soils near forest edges can degrade habitat for certain grape ferns (*Botrychium*).[61]

Pennsylvania forests serve as habitat reserves for many species.[46] Forest fragmentation and loss threaten populations of several breeding birds of conservation concern in Pennsylvania and West Virginia, including wood thrush (*Hylocichla mustelina*), cerulean warbler (*Setophaga cerulea*), and summer tanager (*Piranga rubra*).[62–64] Concern has been raised about potential HVHHF impacts on breeding populations of area-sensitive forest interior songbirds, such as black-throated blue warbler (*Setophaga caerulescens*) and a wide-ranging forest raptor, the northern goshawk (*Accipiter gentilis*).[1] In a 5-year study of breeding birds at 469 sampling points in forest patches ranging from 0.1 to 3,000 ha in Maryland, Pennsylvania, West Virginia, and Virginia, the percentages of forest cover within 2 km and the forest patch area were significant habitat variables for 40 and 38 species, respectively, of 75 species studied; 26 birds were considered area sensitive.[56]

It may take 75–100 years, or more, for cleared forests to regenerate and mature. Forest floor species such as salamanders[65] and herbaceous plants[66] have limited dispersal ability and may take as many additional years to recolonize regrown forests.[67] The guild of forest herbs, often diverse and abundant in mature Appalachian forests, contains many species vulnerable to environmental changes.[66] Logging or

clearing reduces herb diversity, and the herb stratum may take several decades to recover. Herbivory by white-tailed deer (*Odocoileus virginianus*) is harmful to many forest herbs; it is possible that clearing for wellpads, roads, and pipelines may create a landscape that will support more deer and may subject forest herb populations to more intense grazing. One study reported that forests that are less than 70 years old supported fewer rare lichens and bryophytes than older forests;[51] this observation may pertain to young forests that develop following abandonment of HVHHF installations.

Roads and pipelines
Roads act as corridors for the spread of nonnative weeds.[35,68,69] Nonnative or weedy native plants will colonize disturbed soils at roads,[35,70] wellpads, compressor stations, and pipelines, and spread from there into forests and other habitats. This has occurred at energy development sites in western North America.[71] Among possible nonnative weeds that could colonize eastern HVHHF sites are common reed (nonnative haplotype of *Phragmites australis*), stiltgrass (*Microstegium vimineum*), Japanese knotweed (*Polygonum cuspidatum*), spotted knapweed (*Centaurea stoebe*), mugwort (*Artemisia vulgaris*), angelica tree (*Aralia elata*), autumn-olive (*Elaeagnus umbellata*), tree-of-heaven (*Ailanthus altissima*), and empress tree (*Paulownia tomentosa*). These plants thrive on habitats resulting from cut-and-fill, and are colonizing recent disturbances from surface mining, roads, and gas pipelines in the Catskill Mountains and Hudson Highlands of New York and other eastern regions.[72] Common reed disperses along roads, and from there, into adjoining undisturbed habitats,[73,74] where it may adversely affect plant and animal assemblages. The combination of disturbed roadside habitat and salinization from deicing salts is favorable for common reed. Vegetation of pipeline rights-of-way is managed by mowing or spraying herbicide; runoff or spray drift may affect rare native plants in adjoining habitats.

Many forest songbirds avoid roads, trails, pipelines, and human activities.[75] In western Canada, territories of the ovenbird (*Seiurus aurocapillus*) straddled 3-m-wide cleared seismic exploration lines, but did not straddle 8-m-wide lines, leading to local population declines.[75] In another example, red-backed salamander (*Plethodon cinereus*) was less abundant near gravel roads in mature forests in Virginia; this influence of roads on red-backed salamander appeared to be due to dessication of soils.[76] Some access roads and pipelines cross wetlands and streams, potentially creating barriers to movement of water and organisms. It takes an estimated 6,800 truck trips to fracture a single well.[77] Many amphibians, reptiles, birds, and mammals are vulnerable to road mortality; in Ontario, numbers of dead frogs increased, and nearby breeding choruses decreased in intensity, in proportion to the amount of traffic on roads.[78]

Hydrological alteration

Many organisms of streams, wetlands, and temporary ponds require certain patterns of water levels and flows through the year (the hydropattern).[79] Hydrological changes, including the withdrawal of surface waters, and increases in runoff caused by deforestation and impervious surfaces of wellpads and access roads, presumably affect the hydropatterns of streams,[80] floodplains, wetlands, intermittent pools (vernal pools), springs, seeps, shallow groundwater, and karst complexes. Withdrawals from lakes and rivers for fracturing wells might reduce minimum instream flows in the summer. Stream fishes, including brook trout (*Salvelinus fontinalis*), and aquatic invertebrates that must remain in water during summer, such as crayfishes and stoneflies, may be adversely affected by reduced summer flows.[81] Reduced flows may also decrease dissolved oxygen, increase deposition of fine sediment, and increase water temperatures, causing macroinvertebrate species richness to decrease and community composition to shift toward forms tolerant of these conditions.[82] Other species that could potentially be affected include freshwater mussels (Unionoidea), diverse in the Marcellus–Utica region, that are sensitive to hydrology, water quality, and siltation of rivers.[83,84] Hellbender (*Cryptobranchus alleganiensis*), a giant aquatic salamander, requires cool, well-oxygenated, swift streams and is also sensitive to siltation and pollution.[85–87]

In addition, withdrawal and disposal of water could potentially affect groundwater tables and flows, changing groundwater inputs to streams or wetlands. Impacts may be greater during droughts, or where there are competing uses of water, such as in agriculture.[3,13] At a threshold of 10–20% cover by impervious surfaces in a watershed, water quality and species diversity decrease in streams;[80,88–90]

in some HVHHF landscapes, wellpads and access roads cover more than 10%.[1] Because of the density of HVHHF infrastructure on the landscape, and other impacts from siltation and chemical pollution, there may be cumulative impacts to wetlands and streams. Reduction of forest cover in watersheds may also have long-lasting effects on stream biodiversity.[91]

Noise

At HVHHF installations, diesel compressors run 24 h/day, and the noise can be heard from long distances.[20] Continuous loud noise from, for example, transportation networks, motorized recreation, and urban development can interfere with acoustic communication of frogs, birds, and mammals, and cause hearing loss, elevated stress hormone levels, and hypertension in various animals.[92] One study showed that gas compressor station noise in Alberta reduced ovenbird pairing success.[93] In pinyon-juniper woodland of New Mexico, breeding bird species richness was greater, species composition different, and overall nest density similar near gas wellpads without compressors compared to wellpads with compressors, but daily nest survival was higher near pads with compressors due to less predation by western scrub jays (*Aphelocoma californica*).[94] In a comparison of breeding birds near wellpads with and without compressors in the boreal forest, total density and densities of one-third of the individual species were lower at the compressor sites.[95] Bats avoid continuous loud noise and it may impair foraging efficiency.[96–100]

Light

Installations are brightly lit at night,[20] especially wellpads during drilling and fracturing and compressor stations continuously. Artificial night lighting variably affects different taxa; for example, adult moths and aquatic insects may be attracted and killed, whereas various species of bats may be harmed or benefited.[96,101,102] Night lighting potentially disrupts populations of stream insects, in turn affecting food webs and ecosystem function.[103] Mortality, reproduction, and foraging of many other animals are affected negatively or positively.[101] Polarized light pollution from artificial surfaces, especially smoother, darker surfaces including pavement, vehicles, and waste oil, creates another visual disturbance.[104] Animals that orient to polarized light, including many invertebrate and vertebrate taxa, may be killed or have their reproduction disrupted. This potential impact of HVHHF installations has not been studied.

Air quality

Air emissions, including diesel exhaust from compressors and trucks, volatile organics from fracturing fluids, ground-level ozone resulting from their interaction, and road dust, affect air quality around HVHHF sites.[105] Diesel smoke contains mutagenic and carcinogenic polycyclic aromatic hydrocarbons (PAHs)[106] that could affect animal health. In a relevant study, nitrogen oxides from vehicles affected mosses within 50–100 m of roads in England;[107] trees were adversely affected within the same distances, but the haircap moss *Polytrichum commune* showed a decline in frequency with distance from heavily traveled roads.[108] It is possible that diesel exhaust at HVHHF sites could produce similar effects. Lichens are especially sensitive to sulfur dioxide and other air pollutants,[36,39,109,110] and are harmed by road dust, as are sphagnum mosses.[111]

Range-restricted species

A species that has a large part of its geographic range in the Marcellus–Utica region may potentially be at risk of extinction from HVHHF impacts (especially in combination with other widespread environmental change). A recent study[8] analyzed 15 plants, butterflies, fish, amphibians, and mammals with geographic ranges overlapping the Marcellus–Utica region by 36–100% (Figs. 2 and 3). Although most of these species are considered sensitive to forest fragmentation, habitat alteration, or water quality degradation, lungless salamanders (Plethodontidae; eight species analyzed) seemed especially at risk. Many species of invertebrates, higher plants, and cryptogams whose ranges have not been mapped in detail may be quasi-endemic to the region.

Species with larger geographic ranges may nonetheless have important population components or seasonal habitats within the Marcellus–Utica region. The Virginia big-eared bat (*Corynorhinus townsendii virginianus*) occupies 15 limestone caves, 11 of which are in West Virginia.[112] Limestones are often highly porous to water pollution; therefore, cave species could potentially be at greater risk of being affected by HVHHF.

In each state, because of historic, political, social, and economic differences, and genetic differences within many species, environmental impacts on, and management of, rare species differ. Therefore, a species that is restricted to the Marcellus–Utica region within one state could potentially be at higher risk. In Pennsylvania, all known populations of the green salamander (*Aneides aeneus*), and 73% of populations of the snow trillium (*Trillium nivale*), are in localities with a high probability of HVHHF.[1] In New York, bluebreast darter (*Etheostoma camurum*), spotted darter (*E. maculatum*), banded darter (*E. zonale*), and variegate darter (*E. variatum*) are apparently confined to the Marcellus region;[113] these stream fishes are likely to be sensitive to salt and sediment pollution.[114,115]

Species potentially benefiting from HVHHF

Many native organisms use habitats created by construction or abandonment of industrial facilities, such as forest edges or bare soil. Some native bees and wasps dig nest burrows in bare soil, and reptiles often lay eggs in disturbed soils of road and railroad verges. Snakes, including timber rattlesnake (*Crotalus horridus*), are attracted by warm pavement in cooling weather. Several birds nest on bare or sparsely vegetated soil, including mallard (*Anas platyrhynchos*), common nighthawk (*Chordeiles minor*), killdeer (*Charadrius vociferus*), and spotted sandpiper (*Actitis macularia*), and many birds dust bathe on bare soils. White-tailed deer have been shown to be attracted to soils where HVHHF wastewater had been land-applied;[15] porcupine (*Erethizon dorsatum*)[116] and many butterflies[117] would also be attracted to salt. Metal-tolerant vascular plants and mosses could grow in these situations.[118] Postindustrial sites in England are important habitats for beetles, including rare species.[119]

Species of southern affinities would be attracted to wellpads and their peripheries due to solar warming. For example, water-filled wheel ruts and rain pools would serve as larval mosquito habitats; in Wyoming, there was a 75% increase in 5 years in potential mosquito larval habitats in ponds holding wastewater from coal bed gas drilling.[120] Access roads with numerous, long-lasting rain pools might support the globally rare feminine clam shrimp (*Cyzicus gynecia*).[121] It is possible that some grassland and shrubland species might colonize decommissioned facilities if they are extensive or adjoin other nonforested habitats. Most organisms able to colonize active or abandoned installations may be

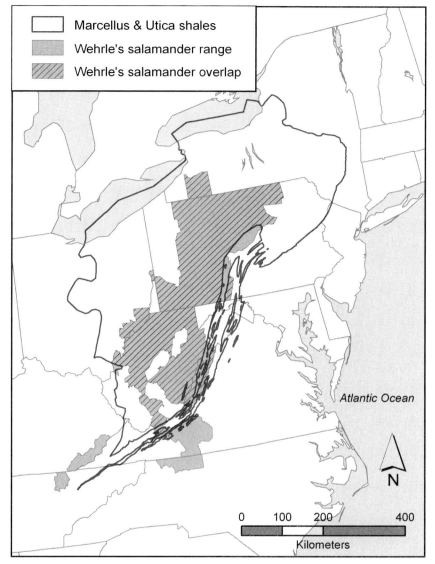

Figure 2. Distribution of Wehrle's salamander (*Plethodon wehrlei*) in relation to the Marcellus–Utica shale region. Reprinted with the permission of Cambridge University Press.[8]

common species and ecological generalists. Rare or sensitive species that are small or require only small habitat patches (e.g., land snails, millipedes, certain insects) may persist in forest patches between well-pads, and some organisms might escape predators or competitors in fragments.

Cumulative impacts
In the Marcellus–Utica region, HVHHF constitutes landscape- and regional-scale activities and impacts.

Many thousands of wellpads will be distributed across the 280,000 km^2 region. Each wellpad will likely be drilled several times, and successful wells will be fractured multiple times during their 40- to 50-year life span.

Widespread environmental changes other than those produced by HVHHF also affect eastern biodiversity,[6,122] including coal mining, logging, agriculture, urban sprawl, accelerated climate change, acidification, eutrophication, chemical

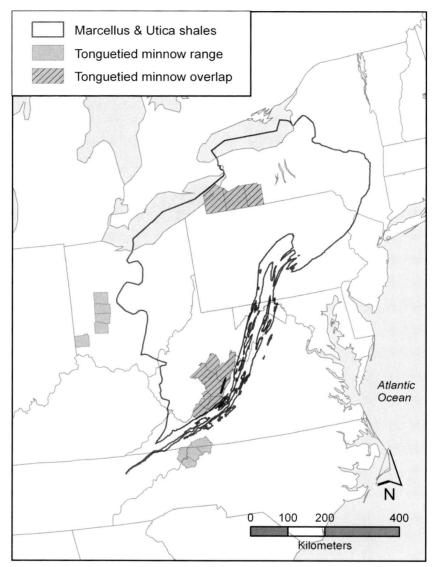

Figure 3. Distribution of tongue-tied minnow (*Exoglossum laurae*) in relation to the Marcellus–Utica shale region. Reprinted with the permission of Cambridge University Press.[8]

contamination, altered fire regimes, emerging pathogens and parasites, and nonnative species spread. For example, most tree species are not shifting latitudinal ranges to keep pace with climate warming, and the ranges of many species are shrinking.[123] Such large-scale changes could potentially interact synergistically with the HVHHF impacts on forest biota as they accumulate across space and time. One study suggested that the effects of HVHHF on stream water quality will accumulate across watersheds.[3] In a meta-analysis of the effects of roads, power lines, and wind turbines on birds and mammals, bird populations were reduced as far as 1 km, and mammal populations were reduced as far as 5 km, from roads and infrastructure.[124] If this finding applies to the wellpads, gas compressors, and roads associated with HVHHF, the corresponding buffers around each installation needed to protect birds and mammals (3.1 km^2 and 78.5 km^2) are larger than the current spacing units for well density in Pennsylvania (1–2.5 km^2) and those projected for New York (2.6 km^2).[16]

Discussion and conclusions

Biodiversity impacts of HVHHF are similar to the impacts of many industries, although the chemical complexity and geographic extent are unusual. The major, long-term effects on biota likely to propagate through landscapes are habitat loss and fragmentation, chemical pollution, degradation of water quality, and hydrological alteration; other impacts, including noise, light, and air quality, may be more local and short-term. Biota vulnerable to HVHHF impacts include many native organisms that are important either for subsistence or in broader markets, such as medicinal plants (e.g., goldenseal (*Hydrastis canadensis*)),[125] edible fungi, brook trout and other sport fishes,[1] game birds and mammals (e.g., wood duck (*Aix sponsa*)), furbearers (American mink (*Mustela vison*), river otter (*Lontra canadensis*), common muskrat (*Ondatra zibethicus*)), and "watchable" wildlife (e.g., many forest-breeding birds). For example, studies suggest that HVHHF may affect trout habitats via water temperature increase, siltation, and heavy metals.[126,127]

Many of the biodiversity impacts of HVHHF might be reduced by zero-loss management of chemicals, wastewater, soil, and other pollutants, but this is a challenge considering the record of leaks, spills, fugitive emissions, and disposal. Water use and truck traffic can be reduced by reusing more wastewater, but similar amounts of pollutants will require disposal. If it eventually becomes possible to drill horizontally several kilometers, fewer wellpads would be needed, thus reducing fragmentation, and allowing more wells to be sited next to highways or on derelict lands, such as abandoned strip mines. However, pipelines would still fragment forests and impinge on sensitive habitats.

Forest loss and fragmentation are considered among the most serious threats to biodiversity.[128,129] Many forest species, particularly birds, require extensive tracts of continuous forest to maintain viable breeding populations. Inasmuch as the eastern United States was extensively deforested during the 1800s, one might ask whether current deforestation and fragmentation matter to biodiversity. At a maximum, only half of the east was deforested at once because clearing was not concurrent across the region; asynchronous deforestation prob-ably prevented extinction of many species.[129] Yet deforestation contributed greatly to the extinction of the passenger pigeon (*Ectopistes migratorius*)[130] and the temporary loss or rarity of red-shouldered hawk (*Buteo lineatus*), wild turkey (*Meleagris gallopavo*), pileated woodpecker (*Dryocopus pileatus*), American beaver, black bear (*Ursus americanus*), fisher (*Martes pennanti*), and white-tailed deer from most of New York State and probably large regions elsewhere in the eastern United States.[131] Most of these species have recovered with the redevelopment of extensive forests, even to the point of overabundance of deer, bear, and turkey. Forest cover in the east is decreasing again,[132] and forests of the conterminous United States are fragmented to the degree that edge effects occur throughout most forested landscapes.[133] Fragmentation also affects grasslands and their breeding birds.[16,134] The many other stressors affecting freshwater organisms[135] may be compounded by water pollution and hydrological alteration from HVHHF.

Biotas are impoverished in industrial and urban areas, although many species thrive, including some rare species.[136–138] Few empirical data are available on biodiversity impacts of eastern HVHHF, although activities are already widespread and potentially will occur throughout 280,000 km^2. HVHHF is also intensive, causing great changes to habitats at HVHHF installations and to the intervening landscapes. Consideration of a broad spectrum of taxa and guilds suggests potential HVHHF risks to biodiversity, particularly organisms that are specialized in their habitat, require unpolluted freshwater with natural hydropatterns, or have small geographic ranges concentrated in the Appalachian Basin. Impoverishment of species assemblages likely will lead to diminution of ecosystem functions and services.[139]

It is expected that an HVHHF installation will be decommissioned and the site restored after 40–50 years; procedures may include regrading, removing roads and impoundments, restoring topsoil, and native planting.[21] Restoration will accomplish more if it is targeted at habitats and species of conservation concern, rather than simply restoring forest or grassland. For example, CWD is important for salamanders, snakes, invertebrates, bryophytes, and lichens. Coarse woody debris could be stockpiled when a site is cleared and used for restoration of a nearby site that is being decommissioned. Construction, operation, and decommissioning of HVHHF facilities, if viewed as a mosaic across the landscape, could be better managed to reduce impacts on biodiversity. Most research on wild organisms is restricted in space and time; thus, we are not well equipped to understand and conserve on large scales.[140] Most regulation of HVHHF has occurred at the level of the individual wellpad; however, to protect biodiversity and ecosystem services, it may be necessary to plan and regulate at the level of the whole Marcellus–Utica region.

Acknowledgments

I am grateful for discussions with Jennifer Gillen, Kristi MacDonald, Cody Mellott, Karen Schneller-McDonald, Gretchen Stevens, and Kristin Westad. The maps were prepared by Maxine Segarnick and Jordan Michael Kincaid, and edited by Kristen Bell Travis, who also edited the manuscript. This work is a Bard College Field Station—Hudsonia contribution.

Conflicts of interest

The author declares no competing financial interest.

References

1. Johnson, N. 2010. Pennsylvania energy impacts assessment. Report 1: Marcellus shale natural gas and wind. Nature Conservancy, Pennsylvania Chapter. Harrisburg, PA. Accessed: September 8, 2012. http://www.nature.org/media/pa/pa_energy_assessment_report.pdf.

2. Colborn, T., C. Kwiatkowski, K. Schultz & M. Bachran. 2011. Natural gas operations from a public health perspective. *Hum. Ecol. Risk Assess.* **17:** 1039–1056.

3. Entrekin, S., M. Evans-White, B. Johnson & E. Hagenbuch. 2011. Rapid expansion of natural gas development poses a threat to surface waters. *Front. Ecol. Environ.* **9:** 503–511.

4. Schmidt, C.W. 2011. Blind rush? Shale gas boom proceeds amid human health questions. *Environ. Health Persp.* **119:** 348–353.

5. Sutherland, W.J., S. Bardsley, L. Bennun, *et al.* 2011. Horizon scan of global conservation issues for 2011. *Trends Ecol. Evol.* **26:** 10–16.

6. Wyman, R.L. 1991. "Multiple threats to wildlife: climate change, acid precipitation, and habitat fragmentation." In *Global Climate Change and Life on Earth.* R.L. Wyman, Ed.: 134–155. New York, NY: Routledge, Chapman and Hall.

7. Highton, R. 2005. "Declines of eastern North American woodland salamanders (*Plethodon*)." In *Amphibian Declines.* M.J. Lannoo, Ed.: 34–46. Berkeley, CA: University of California Press.

8. Gillen, J. & E. Kiviat. 2012. Hydraulic fracturing threats to species with restricted geographic ranges in the eastern United States. *Environ. Pract.* **14:** 320–331. DOI: 10.10170S1466046612000361.

9. Lindenmayer, D.B. & G.E. Likens. 2011. Direct measurement versus surrogate indicator species for evaluating environmental change and biodiversity loss. *Ecosystems* **14:** 47–59.

10. Raphael, M.G. & R. Molina. 2005. *Conservation of Rare or Little-Known Species.* Washington, DC: Island Press.

11. US EIA (U.S. Energy Information Administration). 2012. What is shale gas and why is it important? U.S. Energy Information Administration. Washington, DC. Accessed September 23, 2012. http://www.eia.gov/energy_in_brief/about_shale_gas.cfm.

12. ARI (Advanced Resources International). 2011. World shale gas resources: an initial assessment of 14 regions outside the United States. U.S. Energy Information Administration. Washington, DC. Accessed September 23, 2012. http://www.eia.gov/analysis/studies/worldshalegas/pdf/fullreport.pdf.

13. Soeder, D.J. & W.M. Kappel. 2009. Water resources and natural gas production from the Marcellus shale. U.S. Geological Survey Fact Sheet 2009-3032. Accessed September 16, 2012. http://pubs.usgs.gov/fs/2009/3032/pdf/FS2009-3032.pdf.

14. Bailey, R.G. 1995. *Description of the Ecoregions of the United States,* 2nd ed. Washington, DC: U.S. Forest Service Miscellaneous Publication 1391 (revised).

15. Adams, M.B., M.W. Ford, T.M. Schuler & M. Thomas-Van Gundy. 2011. Effects of natural gas development on forest ecosystems. Proceedings of the 17th Central Hardwood Forest Conference GTR-NRS-P-**78:** 219–226.

16. Davis, J.B. & G.R. Robinson. 2012. A geographic model to assess and limit cumulative ecological degradation from Marcellus shale exploitation in New York, USA. *Ecol. Soc.* **17:** article 25.

17. Collins, E.A. 2010. New withdrawals, new impairments as Pennsylvania develops the Marcellus shale. 2010 Eastern Water Resources Conference. Orlando, FL. May 20–21, 2010. University of Pittsburgh Legal Studies Research Paper 2010–34. Accessed September 19, 2012. http://papers.ssrn.com/sol3/papers.cfm?abstract_id=1699731.

18. Rozell, D.J. & S.J. Reaven. 2012. Water pollution risk associated with natural gas extraction from the Marcellus Shale. *Risk Anal.* **32:** 1382–1393.

19. Michaels, C., J. Simpson & W. Wegner. 2010. Fractured communities: case studies of the environmental impacts of industrial gas drilling. Riverkeeper. Ossining, NY. Accessed September 16, 2012. http://www.riverkeeper.org/wp-content/uploads/2010/09/Fractured-Communities-FINAL-September-2010.pdf.

20. Mellott, C. 2011. Natural gas extraction from the Marcellus formation in Pennsylvania: environmental impacts and possible policy responses for state parks. Annandale, NY: M.S. Thesis, Bard College Center for Environmental Policy.

21. Mitchell, A.L. & E.A. Casman. 2011. Economic incentives and regulatory framework for shale gas well site reclamation in Pennsylvania. *Environ. Sci. Technol.* **45:** 9506–9514.

22. Karraker, N.E. 2008. "Impacts of road deicing salts on amphibians and their habitats." In *Urban Herpetology.* J.C.

Mitchell, R.E.J. Brown & B. Bartholomew, Eds.: 211–223. Salt Lake City, UT: Society for the Study of Amphibians and Reptiles.

23. Nash, T.H. & O.L. Lange. 1988. Responses of lichens to salinity: concentration and time-course relationships and variability among Californian species. *N. Phytol.* **109:** 361–367.

24. Matos, P., J. Cardoso-Vilhena, R. Figueira & A. Jorge Sousa. 2011. Effects of salinity stress on cellular location of elements and photosynthesis in *Ramalina canariensis* Steiner. *Lichenologist* **43:** 155–164.

25. Pisani, T., S. Munzi, L. Paoli, *et al.* 2011. Physiological effects of mercury in the lichens *Cladonia arbuscula* subsp. *mitis* (Sandst.) Ruoss and *Peltigera rufescens* (Weiss) Humb. *Chemosphere* **82:** 1030–1037.

26. Hill M.O., C.D. Preston, S.D.S. Bosanquet & D.B. Roy. 2007. BRYOATT: attributes of British and Irish mosses, liverworts and hornworts. NERC Centre for Ecology and Hydrology and Countryside Council for Wales. Accessed: December 27, 2012. http://nora.nerc.ac.uk/1131/1/BRYOATT.pdf.

27. Wilcox, D.A. 1984. The effects of NaCl deicing salts on *Sphagnum recurvum* P. Beauv. *Environ. Exp. Bot.* **24:** 295–301, 303–304.

28. Wilcox, D.A. & R.E. Andrus. 1987. The role of *Sphagnum fimbriatum* in secondary succession in a road salt impacted bog. *Can. J. Botany.* **65:** 2270–2275.

29. Richburg, J.A., W.A. Patterson III & F. Lowenstein. 2001. Effects of road salt and *Phragmites australis* invasion on the vegetation of a western Massachusetts calcareous lake-basin fen. *Wetlands* **21:** 247–255.

30. Roth, D. & G. Wall. 1976. Environmental effects of highway deicing salts. *Ground Water* **14:** 286-289.

31. Bryson, G.M. & A.V. Barker. 2002. Sodium accumulation in soils and plants along Massachusetts roadsides. *Commun. Soil Sci. Plan.* **33:** 67–78.

32. Smith, M.J., K.M. Ough, M.P. Scroggie, *et al.* 2009. Assessing changes in macrophyte assemblages with salinity in non-riverine wetlands: a Bayesian approach. *Aquat. Bot.* **90:** 137–142.

33. Thouvenot, L., J. Haury & G. Thiébaut. 2012. Responses of two invasive macrophyte species to salt. *Hydrobiologia* **686:** 213–223.

34. Wilcox, D.A. 1986. The effects of deicing salts on vegetation in Pinhook Bog, Indiana. *Can. J. Bot.* **64:** 865–874.

35. Trombulak, S.C. & C.A. Frissell. 2000. Review of ecological effects of roads on terrestrial and aquatic communities. *Conserv. Biol.* **14:** 18–30.

36. Wetmore, C.M. 1988. "Lichen floristics and air quality." In *Lichens, Bryophytes and Air Quality.* (Bibliotheca Lichenologica) T.H. Nash, III & V. Wirth, Eds.: 55–65. Berlin, Germany: J. Cramer.

37. Branquinho, C., D.H. Brown, C. Máguas & F. Catarino. 1997. Lead (Pb) uptake and its effects on membrane integrity and chlorophyll fluorescence in different lichen species. *Environ. Exp. Bot.* **37:** 95–105.

38. Garty, J., Y. Cohen, N. Kloog & A. Karnieli. 1997. Effects of air pollution on cell membrane integrity, spectral reflectance and metal and sulfur concentrations in lichens. *Environ. Toxicol. Chem.* **16:** 1396–1402.

39. Brodo, I.M., S.D. Sharnoff & S. Sharnoff. 2001. *Lichens of North America.* New Haven, CT: Yale University Press.

40. Bačkor, M., B. Pawlik-Skowrońska, J. Tomko, *et al.* 2006. Response to copper stress in aposymbiotically grown lichen mycobiont *Cladonia cristatella*: uptake, viability, ergosterol and production of non-protein thiols. *Mycol. Res.* **110:** 994–999.

41. Heumann, H.G. 1987. Effects of heavy metals on growth and ultrastructure of *Chara vulgaris.* *Protoplasma* **136:** 37–48.

42. Gosek, A., M. Kwiatkowska & R. Duszyński. 1997. The effect of cadmium on growth of vegetative thallus and development of generative organs of *Chara vulgaris* L. after short time of cultivation. *Acta Soc. Bot. Pol.* **66:** 67–72.

43. Bibi M.H., T. Asaeda & E. Azim. 2010. Effects of Cd, Cr, and Zn on growth and metal accumulation in an aquatic macrophyte, *Nitella graciliformis.* *Chem. Ecol.* **26:** 49–56.

44. Long, J.A., T.A. Nelson & M.A. Wulder. 2010. Characterizing forest fragmentation: distinguishing change in composition from configuration. *Appl. Geogr.* **30:** 426–435.

45. Zipperer, W.C. 1993. Deforestation patterns and their effects on forest patches. *Landscape Ecol.* **8:**177–184.

46. Drohan, P.J., M. Brittingham, J. Bishop & K. Yoder. 2012. Early trends in landcover change and forest fragmentation due to shale gas development in Pennsylvania: a potential outcome for the northcentral Appalachians. *Environ. Manage.* **49:** 1061–1075.

47. Demaynadier, P.G. & M.L. Hunter, Jr. 1998. Effects of silvicultural edges on the distribution and abundance of amphibians in Maine. *Conserv. Biol.* **12:** 340–352.

48. Rothermel, B.B. & R.D. Semlitsch. 2006. Consequences of forest fragmentation for juvenile survival in spotted (*Ambystoma maculatum*) and marbled (*Ambystoma opacum*) salamanders. *Can. J. Zool.* **84:** 797–807.

49. Popescu, V.D. & M.L. Hunter, Jr. 2011. Clear-cutting affects habitat connectivity for a forest amphibian by decreasing permeability to juvenile movements. *Ecol. Appl.* **21:** 1283–1295.

50. Moseley, K.R., W.M. Ford & J.W. Edwards. 2009. Local and landscape scale factors influencing edge effects on woodland salamanders. *Environ. Monit. Assess.* **151:** 425–435.

51. Rudolphi, J. & L. Gustafsson. 2011. Forests regenerating after clear-cutting function as habitat for bryophyte and lichen species of conservation concern. *PLoS One* **6:** e18639.

52. Swarts, N.D. & K.W. Dixon. 2009. Terrestrial orchid conservation in the age of extinction. *Ann. Bot.-Lond.* **104:** 543–556.

53. Brunet, J., K. Valtinat, M.L. Mayr, *et al.* 2011. Understory succession in post-agricultural oak forests: habitat fragmentation affects forest specialists and generalists differently. *Forest Ecol. Manag.* **262:** 1863–1871.

54. NYNHP (New York Natural Heritage Program). 2009. Online conservation guide for *Pieris virginiensis.* Accessed: January 9, 2013. http://acris.nynhp.org/guide.php?id=7830.

55. Gibbs, J.P. 1998. Distribution of woodland amphibians along a forest fragmentation gradient. *Landscape Ecol.* **13:** 263–268.

56. Robbins, C.S., D.K. Dawson & B.A. Dowell. 1989. Habitat area requirements of breeding forest birds of the middle Atlantic states. *Wildlife Monogr.* **103:** 3–34.

57. Rolstad, J. 2008. Consequences of forest fragmentation for the dynamics of bird populations: conceptual issues and the evidence. *Biol. J. Linn. Soc.* **42:** 149–163.

58. Smith, A.C., L. Fahrig & C.M. Francis. 2011. Landscape size affects the relative importance of habitat amount, habitat fragmentation, and matrix quality on forest birds. *Ecography* **34:** 103–113.

59. Cox, W.A., F.R. Thompson, B. Root & J. Faaborg. 2012. Declining brown-headed cowbird (*Molothrus ater*) populations are associated with landscape-specific reductions in brood parasitism and increases in songbird productivity. *PloS One* **7:** e47591.

60. McCormick, M.K. & D.F. Whigham. 2012. Using the complexities of orchid life histories to target conservation efforts. *Native Orchid Conf. J.* **9:** 14–23.

61. Johnson-Groh, C.L. & J.M. Lee. 2002. Phenology and demography of two species of *Botrychium* (Ophioglossaceae). *Am. J. Bot.* **89:** 1624–1633.

62. Brauning, D.W., Ed. 1992. *Atlas of Breeding Birds in Pennsylvania.* Pittsburgh, PA: University of Pittsburgh Press.

63. Buckelew, A.R., Jr. & G.A. Hall. 1994. *The West Virginia Breeding Bird Atlas.* Pittsburgh, PA: University of Pittsburgh Press.

64. Steele, M.A., M.C. Brittingham, T.J. Maret & J.F. Merritt. 2010. *Terrestrial Vertebrates of Pennsylvania: A Complete Guide to Species of Conservation Concern.* Baltimore, MD: Johns Hopkins University Press.

65. Green, D.M. 2005. "Biology of amphibian declines." In *Amphibian Declines: The Conservation Status of United States Species.* M. Lannoo, Ed.: 28–33. Berkeley, CA: University of California Press.

66. Jolls, C.L. 2003. "Populations of and threats to rare plants of the herb layer." In *The Herbaceous Layer in Forests of Eastern North America.* F.S. Gilliam & M.R. Roberts, Eds.:105–159. Oxford, UK: Oxford University Press.

67. Bratton, S.P. & A.J. Meier. 1998. Restoring wildflowers and salamanders in southeastern deciduous forests. *Restor. Manage. Notes* **16:** 158–165.

68. Thiele, J., U. Schuckert & A. Otte. 2008. Cultural landscapes of Germany are patch-corridor-matrix mosaics for an invasive megaforb. *Landscape Ecol.* **23:** 453–465.

69. Mortensen, D.A., E.S.J. Rauschert, A.N. Nord & B.P. Jones. 2009. Forest roads facilitate the spread of invasive plants. *Invas. Plant Sci. Manage.* **2:** 191–199.

70. Forman, R.T.T. & L.E. Alexander. 1998. Roads and their major ecological effects. *Annu. Rev. Ecol. Syst.* **29:** 207–231.

71. Evangelista, P.H., A.W. Crall & E. Bergquist. 2011. "Invasive plants and their response to energy development." In *Energy Development and Wildlife Conservation in Western North America.* D.E. Naugle, Ed.: 115–129. Washington, DC: Island Press.

72. Kiviat, E. unpublished data.

73. Catling, P.M. & S. Carbyn. 2006. Recent invasion, current status and invasion pathway of European common reed, *Phragmites australis* subspecies *australis*, in the southern Ottawa district. *Can. Field Nat.* **120:** 307–312.

74. Jodoin, Y., C. Lavoie, P. Villeneuve, *et al.* 2008. Highways as corridors and habitats for the invasive common reed *Phragmites australis* in Quebec, Canada. *J. Appl. Ecol.* **45:** 459–466.

75. Bayne, E.M. & B.C. Dale. 2011. "Effects of energy development on songbirds." In *Energy Development and Wildlife Conservation in Western North America.* D.E. Naugle, Ed.: 95–114. Washington, DC: Island Press.

76. Marsh, D.M. & N.G. Beckman. 2004. Effects of forest roads on the abundance and activity of terrestrial salamanders. *Ecol. Appl.* **14:** 1882–1891.

77. Garti, A.M. 2012. The illusion of the blue flame: water law and unconventional gas drilling in New York State. *Environ. Law N.Y.* **23:** 159–165.

78. Fahrig, L., J.H. Pedlar, S.E. Pope, *et al.* 1995. Effect of road traffic on amphibian density. *Biol. Conserv.* **73:** 177–182.

79. Batzer, D.P. & R.R. Sharitz. 2006. *Ecology of Freshwater and Estuarine Wetlands.* Berkeley, CA: University of California Press.

80. Klein, R.D. 1979. Urbanization and stream quality impairment. *Water Resour. Bull.* **15:** 948–963.

81. Orth, D.J. & O.E. Maughan. 1982. Evaluation of the incremental methodology for recommending instream flows for fishes. *T. Am. Fish. Soc.* **111:** 413–445.

82. Dewson, Z.S., A.B.W. James & R.G. Death. 2007. A review of the consequences of decreased flow for instream habitat and macroinvertebrates. *J. N. Am. Benthol. Soc.* **26:** 401–415.

83. Lydeard, C., R.H. Cowie, W.F. Ponder, *et al.* 2004. The global decline of nonmarine mollusks. *BioScience* **54:** 321–330.

84. Baldigo, B.P., A.G. Ernst, G.E. Schuler & C.D. Apse. 2008. Relations of environmental factors with mussel-species richness in the Neversink River, New York. U.S. Geological Survey Open-File Report 2007–1283.

85. Williams, R.D., J.E. Gates, C.E. Hocutt & G.J. Taylor. 1981. The hellbender: a nongame species in need of management. *Wildlife Soc. B.* **9:** 94–100.

86. Nickerson, M.A., K.L. Krysko & R.D. Owen. 2002. Ecological status of the hellbender (*Cryptobranchus alleganiensis*) and the mudpuppy (*Necturus maculosus*) salamanders in the Great Smoky Mountains. *J. N.C. Acad. Sci.* **118:** 27–34.

87. Hopkins, W.A. & S.E. DuRant. 2011. Innate immunity and stress physiology of eastern hellbenders (*Cryptobranchus alleganiensis*) from two stream reaches with differing habitat quality. *Gen. Comp. Endocr.* **174:** 107–115.

88. Schueler, T.R. 1994. The importance of imperviousness. *Watershed Protect. Tech.* **1:** 100–111.

89. Arnold, C.L. & C.J. Gibbons. 1996. Impervious surface coverage: the emergence of a key environmental indicator. *J. Am. Plann. Assoc.* **62:** 243–258.

90. NRC (National Research Council). 2008. *Urban Stormwater Management in the United States.* Washington, DC: National Academies Press.

91. Harding, J.S., E.F. Benfield, P.V. Bolstad., *et al.* 1998. Stream biodiversity: the ghost of land use past. *Proc. Natl. Acad. Sci. USA* **95:** 14843–14847.

92. Barber, J.R., K.R. Crooks & K.M. Fristrup. 2010. The costs of chronic noise exposure for terrestrial organisms. *Trends Ecol. Evol.* **25:**180–189.

93. Habib, L., E.M. Bayne & S. Boutin. 2007. Chronic industrial noise affects pairing success and age structure of ovenbirds *Seiurus aurocapilla*. *J. Appl. Ecol.* **44:** 176–184.

94. Francis C.D., C.P. Ortega & A. Cruz. 2009. Noise pollution changes avian communities and species interactions. *Curr. Biol.* **19:** 1415–1419.

95. Bayne, E.M., L. Habib & S. Boutin. 2008. Impacts of chronic anthropogenic noise from energy-sector activity on abundance of songbirds in the boreal forest. *Conserv. Biol.* **22:** 1186–1193.

96. Legakis, A., C. Papadimitriou, M. Gaethlich & D. Lazaris. 2000. Survey of the bats of the Athens metropolitan area. *Myotis* **38:** 41–46.

97. Jones, G. 2008. Sensory ecology: noise annoys foraging bats. *Curr. Biol.* **18:** 1098–1100.

98. Schaub, A., J. Ostwald & B.M. Siemers. 2008. Foraging bats avoid noise. *J. Exp. Biol.* **211:** 3174–3180.

99. Siemers, B.M. & A. Schaub. 2011. Hunting at the highway: traffic noise reduces foraging efficiency in acoustic predators. *Proc. R. Soc. B.-Biol. Sci.* **278:** 1646–1652.

100. Francis, C.D., N.J. Kleist, C.P. Ortega & A. Cruz. 2012. Noise pollution alters ecological services: enhanced pollination and disrupted seed dispersal. *Proc. R. Soc. B.-Biol. Sci.* **279:** 2727–2735.

101. Rich, C. & T. Longcore, Eds.: 2006. *Ecological Consequences of Artificial Night Lighting*. Covelo, CA: Island Press.

102. Stone, E.L., G Jones & S. Harris. 2009. Street lighting disturbs commuting bats. *Curr. Biol.* **19:** 1123–1127.

103. Perkin, E.K., F. Hölker, J.S. Richardson, *et al.* 2011. The influence of artificial light on stream and riparian ecosystems: questions, challenges, and perspectives. *Ecosphere* **2:** 1–16.

104. Horváth, G., G. Kriska, P. Malik & B. Robertson. 2009. Polarized light pollution: a new kind of ecological photopollution. *Front. Ecol. Environ.* **7:** 317–325.

105. Colborn, T., K. Schultz, L. Herrick & C. Kwiatkowski. 2012. An exploratory study of air quality near natural gas operations. *Hum. Ecol. Risk Assess.* DOI: 10.1080/10807039.2012.749447.

106. Courter, L.A., C. Pereira & W.A. Baird. 2007. Diesel exhaust influences carcinogenic PAH-induced genotoxicity and gene expression in human breast epithelial cells in culture. *Mutat. Res.* **625:** 72–82.

107. Bignal, K.L., M.R. Ashmore & A.D. Headley. 2008. Effects of air pollution from road transport on growth and physiology of six transplanted bryophyte species. *Environ. Pollut.* **156:** 332–340.

108. Bignal, K.L., M.R. Ashmore, A.D. Headley, *et al.* 2007. Ecological impacts of air pollution from road transport on local vegetation. *Appl. Geochem.* **22:** 1265–1271.

109. Seaward, M.R.D. 1993. Lichens and sulphur dioxide air pollution: field studies. *Environ. Rev.* **1:** 73–91.

110. Lisowska, M. 2011. Lichen recolonisation in an urban-industrial area of southern Poland as a result of air quality improvement. *Environ. Monit. Assess.* **179:** 177–190.

111. Farmer, A.M. 1993. The effects of dust on vegetation—a review. *Environ. Pollut.* **79:** 63–75.

112. Wilcove, D., S. Roble & G. Hammerson. 2011. Virginia big-eared bat *(Corynorhinus townsendii virginianus)*. Nature-Serve Explorer. An online encyclopedia of life. Version 7.1. NatureServe. Arlington, VA. Accessed September 17, 2012. http://www.natureserve.org/explorer/servlet/NatureServe?searchName=Corynorhinus+townsendii+virginianus.

113. Smith, C.L. 1985. *The Inland Fishes of New York State*. Albany, NY: New York State Department of Environmental Conservation.

114. Cañedo-Argüelles, M., B.J. Kefford, C. Piscart, *et al.* 2013. Salinisation of rivers: an urgent ecological issue. *Environ. Pollut.* **173:** 157–167.

115. NatureServe. 2013. NatureServe Explorer: an online encyclopedia of life [web application]. Version 7.1. NatureServe, Arlington, Virginia. Accessed April 11, 2013. http://www.natureserve.org/explorer.

116. Roze, U. 1989. *The North American Porcupine*. Washington, D.C: Smithsonian Institution Press, 261.

117. Molleman, F. 2009. Puddling: from natural history to understanding how it affects fitness. *Entomol. Exp. Appl.* **134:** 107–113.

118. Salemaa, M., I. Vanha-Majamaa & J. Derome. 2001. Understorey vegetation along a heavy-metal pollution gradient in SW Finland. *Environ. Pollut.* **112:** 339–350.

119. Eyre, M.D., M.L. Luff & J.C. Woodward. 2003. Beetles (Coleoptera) on brownfield sites in England: an important conservation resource? *J. Insect Conserv.* **7:** 223–231.

120. Zou, L., S.N. Miller & E.T. Schmidtmann. 2006. Mosquito larval habitat mapping using remote sensing and GIS: implications of coalbed methane development and West Nile virus. *J. Med. Entomol.* **43:** 1034–1041.

121. Schmidt, R.E. & E. Kiviat. 2007. State records and habitat of clam shrimp, *Caenestheriella gynecia* (Crustacea: Conchostraca), in New York and New Jersey. *Can. Field-Naturalist* **121:** 128–132.

122. Root, T.L., J.T. Price, K.R. Hall, S.H. Schneider, *et al.* 2003. Fingerprints of global warming on wild animals and plants. *Nature* **421:** 57–60.

123. Zhu, K., C.W. Woodall & J.S. Clark. 2012. Failure to migrate: lack of tree range expansion in response to climate change. *Glob. Change Biol.* **18:** 1042–1052.

124. Benítez-López, A., R. Alkemade & P.A. Verweij. 2010. The impacts of roads and other infrastructure on mammal and bird populations: a meta-analysis. *Biol. Conserv.* **143:** 1307–1316.

125. Sanders, S.E. & J.B. McGraw. 2005. Harvest recovery of goldenseal, *Hydrastis canadensis* L. *Am. Midl. Nat.* **153:** 87–94.

126. Harper, D.D., A.M. Farag, C. Hogstrand & E. MacConnell. 2009. Trout density and health in a stream with variable water temperatures and trace element concentrations: does a cold-water source attract trout to increased metal exposure? *Environ. Toxicol. Chem.* **28:** 800–808.

127. Siitari, K.J., W.W. Taylor, S.A.C. Nelson & K.E. Weaver. 2011. The influence of land cover composition and groundwater on thermal habitat availability for brook charr (*Salvelinus fontinalis*) populations in the United States of America. *Ecol. Freshw. Fish.* **20:** 431–437.

128. Wilcox, B.A. & D.D. Murphy. 1985. Conservation strategy: the effects of fragmentation on extinction. *Am. Nat.* **125:** 879–887.

129. Pimm, S.L. & R.A. Askins. 1995. Forest losses predict bird extinctions in eastern North America. *Proc. Natl. Acad. Sci. USA* **92:** 9343–9347.

130. Blockstein, D.E. 2002. "Passenger pigeon *(Ectopistes migratorius)*." In *The Birds of North America*. A. Poole & F. Gill, Eds.: 611. Philadelphia, PA: Academy of Natural Sciences.

131. Foster, D.R., G. Motzkin, D. Bernardos & J. Cardoza. 2002. Wildlife dynamics in the changing New England landscape. *J. Biogeogr.* **29:** 1337–1357.

132. Drummond, M.A. & T.R. Loveland. 2010. Land-use pressure and a transition to forest cover loss in the eastern US. *BioScience* **60:** 286–298.

133. Riitters, K.H., J.D. Wickham, R.V. O'Neill, *et al.* 2002. Fragmentation of continental United States forests. *Ecosystems* **5:** 815–822.

134. Ribic, C.A., R.R. Koford, J.R. Herkert, *et al.* 2009. Area sensitivity in North American grassland birds: patterns and processes. *Auk* **126:** 233–244.

135. Strayer, D. 2006. Challenges for freshwater invertebrate conservation. *J. N. Am. Benthol. Soc.* **25:** 271–287.

136. Gilbert, O.L. 1989. *The Ecology of Urban Habitats*. London, UK: Chapman & Hall.

137. Savard, J.-P.L., P. Clergeau & G. Mennechez. 2000. Biodiversity concepts and urban ecosystems. *Landscape Urban Plan.* **48:** 131–142.

138. Kiviat, E. & K. MacDonald. 2004. Biodiversity patterns and conservation in the Hackensack Meadowlands, New Jersey. *Urban Habitats* **2:** 28–61.

139. Loreau, M. 2010. Linking biodiversity and ecosystems: towards a unifying ecological theory. *Philos. Trans. R. Soc. B.* **365:** 49–60.

140. May, R.M. 1994. "The effects of spatial scale on ecological questions and answers." In *Large-scale Ecology and Conservation Biology*. P.J. Edwards, R.M. May & N.R. Webb, Eds.: 1–17. Oxford, U.K: Blackwell Science.

Ann. N.Y. Acad. Sci. ISSN 0077-8923

ANNALS OF THE NEW YORK ACADEMY OF SCIENCES
Issue: *The Year in Ecology and Conservation Biology*

Translocation of imperiled species under changing climates

Mark W. Schwartz[1] and Tara G. Martin[2]

[1]John Muir Institute of the Environment, University of California, Davis, California. [2]Climate Adaptation Flagship, CSIRO Ecosystem Sciences, Ecoscience Precinct, Queensland, Australia

Address for correspondence: Mark W. Schwartz, Department of Environmental Science & Policy, 1 Shields Avenue, University of California, Davis, CA 95616. mwschwartz@ucdavis.edu

Conservation translocation of species varies from restoring historic populations to managing the relocation of imperiled species to new locations. We review the literature in three areas—translocation, managed relocation, and conservation decision making—to inform conservation translocation under changing climates. First, climate change increases the potential for conflict over both the efficacy and the acceptability of conservation translocation. The emerging literature on managed relocation highlights this discourse. Second, conservation translocation works in concert with other strategies. The emerging literature in structured decision making provides a framework for prioritizing conservation actions—considering many possible alternatives that are evaluated based on expected benefit, risk, and social–political feasibility. Finally, the translocation literature has historically been primarily concerned with risks associated with the target species. In contrast, the managed relocation literature raises concerns about the ecological risk to the recipient ecosystem. Engaging in a structured decision process that explicitly focuses on stakeholder engagement, problem definition and specification of goals from the outset will allow creative solutions to be developed and evaluated based on their expected effectiveness.

Keywords: conservation; managed relocation; structured decision making; translocation; risk

Introduction

Habitat loss, invasive species, climate change, and other drivers of ecosystem change are resulting in increasing rates of species imperilment and decreasing success of traditional *in situ* conservation methods.[1] The failing capacity to conserve biodiversity is driving a suite of more radical approaches.[1–12] Zoo and botanic garden managers are rethinking their organizational missions;[13–17] protected area managers, their objectives;[4,7,18–20] and endangered species managers their strategies.[21–25] Within this spectrum, translocation of imperiled species is increasingly used as a conservation strategy.[26–28] However, moving species to conserve them has its own challenges. Translocation is a process that often fails,[29–33] moving species can lead to conflict among human interests in conservation,[34,35] translocation may cause unintended negative consequences to either the target species[36] or the re-

cipient ecosystem, and finally translocation can be costly relative to other actions.[37] Under scenarios of climate change the challenges of translocation are exacerbated. For example, conservation translocations under changed climates may need to occur outside of historical distributions, referred to as managed relocation of species, and this raises additional complexities in an already challenging process.

Our purpose is to review the recent literature on conservation translocations and decision making in conservation to synthesize translocation recommendations and to focus future scientific study. Managed relocation, the conservation translocation of species outside historic distributions in anticipation of changing future climates, generates significant scientific and public concern and requires a formal decision process to evaluate the potential benefits and risks. We are just beginning to experience the ecological impacts of anthropogenic

doi: 10.1111/nyas.12050

climate change, with the severest projected impacts on natural ecosystems yet to come.[38] Thus, there is time to plan—to develop adaptation strategies and protocols to help conserve biodiversity under climate change. With significant uncertainty about how climate will change, as well as how ecosystems will respond to the myriad other drivers of future environmental change, action comes with a large risk of making management mistakes.[39,40] Likewise, failing to act, or to act in a timely manner ultimately risks species extinction.[41] Risk generates conflict and controversy over the appropriate steps on behalf of conservation.

There are numerous cases where the need for conservation translocation is immediate.[27,28,42] These cases provide opportunities to work out protocols and processes to minimize conflict and maximize the opportunity for success. Further, there is growing recognition that dramatic, sometimes unprecedented and controversial, management actions will be required in order to maintain biological diversity through this century and into the next.[7,24,43,44] As a consequence, there is a rapidly growing research interest in understanding the trigger points for engaging in novel management actions.[26]

The conservation translocation literature is robust with examples. As climate change drives new conservation thinking, the emerging literature on managed relocation is adding new insights to this rich literature.[45–47] We integrate emerging literature on (1) imperiled species translocation, (2) managed relocation, and (3) conservation decision making to better understand how managers may plan for climate change adaptations that include translocation of imperiled species. Specifically, we focus on the suite of strategies directed at reducing extinction risk by managing the distribution of imperiled species in natural habitats. In treating this subject, we call specific attention to three issues often overlooked in translocation efforts. First, we focus on biological assessments and planning within a decision support framework to foster careful evaluation of translocation against alternative management options. Second, we draw attention to the requirement of risk assemement, including the risk of negative ecological impacts of translocation on recipient ecosystems. Finally, we highlight the need for translocation plans to formally integrate stakeholder concerns and social acceptability into all phases of the management action. As future transloc-

cations may increasingly occur outside historic distributions, it becomes critically important to use a thoughtful and structured process to integrate social and scientific concerns into planning, monitoring, and assessing impacts.

Conservation translocations span a continuum from modest population augmentation to establishing populations in locations with no history of previous occupancy.[23] Terminology to describe these actions, however, is variable.[48] We restrict our usage of this terminology to conservation applications. We use the term *conservation translocation* for any action that involves moving individuals of a species from one location to another for a conservation purpose (e.g., extinction risk abatement, population restoration, establishment in predator-free habitats, reinforcing low populations, increasing genetic variability, improving ecological function).[26] We use the term *managed relocation* to refer to that subset of conservation translocations that are restricted to moving species outside their historic distributions for the purpose of establishing and managing populations in response to changing climate.[48] *Assisted migration* and *assisted colonization* are generally considered synonyms of managed relocation.[47,49–51] We prefer *managed relocation* because it mandates a postrelease management obligation, whereas the more commonly used *assisted colonization* does not.[47]

With the recent explosion of literature on species translocations generally and managed relocation specifically, we attempt to clarify critical research questions to facilitate sound conservation translocation decisions that integrate scientific understanding and social concerns. The managed relocation literature has exploded; over 75% of 195 papers (ISI search, 15 June, 2012 on title words *managed relocation* or *assisted colonization* or *assisted migration*) have been published since 2008. This literature has ranged from empirical descriptions of projects,[52] to rationale for decision support[46] and debate regarding the scientific,[24,53–57] ethical,[11,58–60] and legal[3,61–63] issues associated with managed relocation. In contrast, the conservation translocation literature is robust, with numerous case studies emerging each year.[27,28]

Finally, an array of applied tools, variously called structured decision making, decision analysis, or decision support, have emerged from decision theory and multiattribute utility theory[64,65] and are

Table 3. A hypothetical imperiled species translocation planning timeline to develop decision support

Phase II: Project planning	
Decision analysis	**Sources, tools, frameworks**
Describe problem, objectives, engage stakeholders, define alternative actions, consequences, and trade-offs	Structured decision making[71,94]
Spatial planning	
Assessing potential translocation sites	Species distribution modeling; patch occupancy models[133,134]
Finding an efficient spatial solution	Spatial PVA's combined with Marxan or zonation
Assessing connectedness among population and proposed sites	dispersal modeling[135]
Risk assessment	
Target species risk	Populations genetics, behavioral ecology, disease risk[91,92]
Recipient ecosystem risk	Modeling ecosystem impacts; develop a termination strategy
Institutional procedures	
Permissions	Obtain necessary permits; assess cross-institutional needs
Cooperation	Develop a project team;[105,106] develop a stakeholder network

majority of translocations are conducted without an obvious integrated biological assessment.[91] Among proposed planning requirements, three overarching goals related to harm repeatedly emerge. The proposed action must strive to harm neither (1) the extant population; (2) the individuals being translocated; nor (3) the recipient ecosystem. This assessment of harm includes issues associated with disease transmission within the target species and between the target and other species.[92,93] These concerns also include genetic issues for both the existing and new populations.[26] With respect to harming recipient ecosystems, translocations to areas where the target has been extirpated generally involve the assumption that the recipient ecosystem can absorb the species without negative impact. This assumption comes into question for managed relocation where species are being introduced to new ecosystems. Naturally, decisions often must consider offsetting harm, and therein lies the challenge of making decisions regarding managed relocation.

Through the application of a structured decision making process,[71,94] managers can plan by evaluating the options given by the objectives of the translocation program and the consequences of alternative management strategies. This careful evaluation of

alternative strategies is essential for making a robust case for a translocation effort.

After being satisfied that individuals of the target species can be feasibly moved with minimal risk to the target species, recipient locations must be identified (Table 3). A solid working hypothesis for the habitat requirements of the target species is needed. In the case of climate change, this must include an assessment of the capacity of the species to persist within its historic distribution. Species distribution modeling or habitat occupancy modeling can be used to project this suitable habitat.[95,96] Combined with information on species dynamics, species distribution models can provide information on population viability.[97]

With respect to concerns for adverse impacts to either target species or recipient ecosystems, risk assessment frameworks may be appropriate for understanding the potential negative impacts of translocation efforts (Table 3). Most efforts to summarize translocation risk focus on the risk to the target species.[26] Elements to consider include the conservation status of the site, tenure security, and minimization of opportunities for hybridization and invasiveness.[78] Several types of sites are recommended for exclusion: sites of high species endemism, IUCN category 1 reference reserves, and fully functional threatened ecological communities.[78] Humans have a long history of intentional and unintentional alteration of species distributions.[98] Hubris with respect to tinkering with the distributions of species costs millions of dollars in management costs each year[99] and drives untold ecological damage.[98] Even though there have been few examples of translocated threatened species creating adverse ecological impacts, many ecologists remain concerned.[47,53,55,57,100,101] As climate change adaptation strategies suggest more dramatic translocations, the risk of adverse impacts from our conservation efforts increases, demanding that more attention is paid to risk assessment.

Risk assessment of adverse impacts to recipient ecosystems should include an appraisal of termination costs—is it feasible and what might it cost to control or eradicate the target translocated species should it prove to have negative impacts on the recipient ecosystem? If there are no good eradication methods possible (e.g., many insect taxa), then more effort is warranted to provide assurances that the risk of adverse impacts is minimal. The consequence is that we may, in the end, choose to forego translocation efforts on behalf of species, despite compelling evidence of a possible extinction, simply because the consequences of adverse impacts are too great. There may be many cases where these risks suggest increased emphasis on less risky (lower cost) but potentially less successful (lower benefit) and more socially acceptable (more feasible) strategies like augmentation of existing populations (Table 3).

Institutional procedures may create barriers to conservation action involving translocations (Table 3). There may need to be changes to existing environmental legislation to facilitate managed relocation.[78] Permits for collection, husbandry, transportation, handling, and release may all be required for targets of translocation. Interagency or even international agreements may be required. Changes to national laws or international guidelines may also be needed.[48] For example, under recent IUCN guidelines, populations introduced outside the natural range of the taxon are not assessed globally as contributing to the conservation status of a species.[102] Under this rule, managed relocation may not make a positive contribution to IUCN Red List conservation status[103] or to the Millennium Assessment Goals.[102]

A critical function of a manager is consideration of the ethical and social acceptability of a translocation action among stakeholders (Fig. 1). Many notable translocation cases have generated public controversy (e.g., bears and wolves[34,35,83]). With future climate change, more conservation translocations may be managed relocations, which carry their own potential for controversy.[53,54,57,104] Systematic conservation planning,[105,106] the Open Standards Framework for the Practice of Conservation,[73] and structured decision making[71] each provide a framework for engaging stakeholders with different views.

With social discord over a proposed translocation, the identity of the manager or decision maker is important. For example, if the decision maker is a government agency, then engaging private stakeholders is often a requirement, but private stakeholders may have a limited role in the decision process.[3,48] In contrast, private individuals may be viewed as advocates for a specialized point of view that may not be universally supported. The Torreya Guardians, as an example, translocated an endangered tree species onto private lands over 500 km outside the species' known historic distribution.[107] This was a private action that appears legal despite the target species, *Torreya taxifolia*, being regulated under the U. S. Endangered Species Act.[48] Given that this is a conservation translocation into a new biogeographical location (managed relocation) with no apparent plan to evaluate potential adverse ecosystem impacts or exert population control of the species if there is spread, one could also argue that there is an ethical obligation to public interest.[11,59,62]

Ethicists have described differences in ethical frameworks for conservation related to the combined responsibilities for positive and negative duties to nature:[11,48] a drive to do good, but also to not do harm. Managed relocation has the potential to do both good and harm with respect to socially defined biodiversity values. Stakeholders can become entrenched in their own worldview and it is quite likely that coalitions will form to advocate on behalf of benefits to the target species as well as risks to the recipient ecosystems.[108,109] Resolutions, where they are made, focus on building trust through careful negotiation and collaborative decision making.[110–112] Each of the conservation frameworks discussed here

70. Lunt, I.D. *et al.* 2013. Using assisted colonisation to conserve biodiversity and restore ecosystem function under climate change. *Biol. Conserv.* **157:** 172–177.

71. Gregory, R. *et al.* 2012. *Structured Decision Making: A Practical Guide to Environmental Management Choices.* Wiley Blackwell. Chichester.

72. Schwartz, M.W. *et al.* 2012. Perspectives on the open standards for the practice of conservation. *Biol. Conserv.* **155:** 169–177.

73. CMP. 2007. Open standards for the practice of conservation. Version 2.0., Vol. 2012. Conservation Measures Partnership.

74. Martin, J. *et al.* 2009. Structured decision making as a conceptual framework to identify thresholds for conservation and management. *Ecol. Appl.* **19:** 1079–1090.

75. Kindall, J.L. *et al.* 2011. Population viability analysis to identify management priorities for reintroduced Elk in the Cumberland Mountains, Tennessee. *J. Wildl. Manage.* **75:** 1745–1752.

76. Glick, P.B., B.A. Stein & N.A. Edelson. 2011. *Scanning the Conservation Horizon: A Guide to Climate Change Vulnerability Assessment.* National Wildlife Federation. Washington, DC.

77. Mkanda, F.X. 1996. Potential impacts of future climate change on nyala Tragelaphus angasi in Lengwe National Park, Malawi. *Clim. Res.* **6:** 157–164.

78. Burbidge, A.A. *et al.* 2011. Is Australia ready for assisted colonisation? Policy changes required to facilitate translocations under climate change. *Pac. Conserv. Biol.* **17:** 259–269.

79. Susskind, L., A.E. Camacho & T. Schenk. 2012. A critical assessment of collaborative adaptive management in practice. *J. Appl. Ecol.* **49:** 47–51.

80. Bottrill, M.C. *et al.* 2008. Is conservation triage just smart decision making? *Trends Ecol. Evol.* **23:** 649–654.

81. McDonald-Madden, E. *et al.* 2011. Allocating conservation resources between areas where persistence of a species is uncertain. *Ecol. Appl.* **21:** 844–858.

82. Murdoch, W. *et al.* 2007. Maximizing return on investment in conservation. *Biol. Conserv.* **139:** 375–388.

83. Majic, A. *et al.* 2011. Dynamics of public attitudes toward bears and the role of bear hunting in Croatia. *Biol. Conserv.* **144:** 3018–3027.

84. Martin, T.G. *et al.* 2007. Optimal conservation of migratory species. *PLoS One* **2:** e751.

85. Chadès, I. *et al.* 2011. General rules for managing and surveying networks of pests, diseases and endangered species. *Proc. Natl. Acad. Sci.* **108:** 8323–8328.

86. Runge, M.C., S.J. Converse & Lyons, J.E. 2011. Which uncertainty? Using expert elicitation and expected value of information to design an adaptive program. *Biol. Conserv.* **144:** 1214–1223.

87. Chadès, I., R. Sabbadin, J. Carwardine, *et al.* 2012. MOMDPs: a solution for modelling adaptive management problems. Twenty-Sixth AAAI Conference of the Association for the Advancement of Artificial Intelligence (www.aaai.org). (AAAI-12) July 22-26, Toronto, Canada.

88. Chadès, I., R. Sabbadin, J. Carwardine, *et al.* 2012. MOMDPs: a solution for modelling adaptive management problems. Twenty-Sixth AAAI Conference of the Association for the Advancement of Artificial Intelligence (www.aaai.org). (AAAI-12) July 22-26, Toronto, Canada.

89. Chauvenet, A.L.M. *et al.* 2012. Maximizing the success of assisted colonizations. *Ani. Conserv.* DOI: 10.1111/j.1469-1795.2012.00589.x.

90. IUCN. 1987. IUCN position statement on the translocation of living organisms. Species Survival Commission, C. o. E. a. t. C. o. E.P., Law and Administration, Ed. International Union for Conservation of Nature. Gland, Switzerland.

91. Sheean, V.A., A.D. Manning & D.B. Lindenmayer. 2012. An assessment of scientific approaches towards species relocations in Australia. *Austral Ecol.* **37:** 204–215.

92. Sainsbury, A.W. & R.J. Vaughan-Higgins. 2012. Analyzing disease risks associated with translocations. *Conserv. Biol.* **26:** 442–452.

93. Sainsbury, A.W., D.P. Armstrong & J.G. Ewen. 2012. Methods of disease risk analysis for reintroductions. In *Reintroduction Biology: Integrating Science and Management.* J.G. Ewen *et al.*, Eds.: Wiley Blackwell. Chichester, UK.

94. Hammond, J.S., R.L. Keeney & H. Faiffa. 1999. *Smart Choices: A Practical Guide to Making Better Decisions.* Harvard Business School Press. Boston, MA.

95. Armstrong, D.P. & M.H. Reynolds. 2012. Modelling reintroduction populations: the state of the art and future directions. In *Reintroduction Biology: Integrating Science and Management.* J.G. Ewen *et al.*, Eds.: 165–222. Wiley Blackwell. Chichester, UK.

96. Osborne, P.E. & P.J. Seddon. 2012. Selecting suitable habitat for reintroductions: variation, change and the role of species distribution modelling. In *Reintroduction Biology: Integrating Science and Management.* J.G. Ewen *et al.*, Eds.: 73–104. Wiley Blackwell. Chichester, UK.

97. Fordham, D.A. *et al.* 2012. Plant extinction risk under climate change: are forecast range shifts alone a good indicator of species vulnerability to global warming? *Glob. Change Biol.* **18:** 1357–1371.

98. Mack, R.N. *et al.* 2000. Biotic invasions: causes, epidemiology, global consequences, and control. *Ecol. Appl.* **10:** 689–710.

99. Pimentel, D., R. Zuniga & D. Morrison. 2005. Update on the environmental and economic costs associated with alien-invasive species in the United States. *Ecol. Econ.* **52:** 273–288.

100. Kreyling, J. *et al.* 2011. Assisted colonization: a question of focal units and recipient localities. *Restor. Ecol.* **19:** 433–440.

101. Ricciardi, A. & D. Simberloff. 2009. Assisted colonization: good intentions and dubious risk assessment. *Trends Ecol. Evol.* **24:** 476–477.

102. United Nations. 2010. The Millennium Development Goals Report 2010 United Nations. New York.

103. Butchart, S.H.M. *et al.* 2007. Improvements to the red list index. *PLoS One* **2:** e140.

104. Aubin, I. *et al.* 2011. Why we disagree about assisted migration: ethical implications of a key debate regarding the future of Canada's forests. *Forest Chron.* **87:** 755–765.

105. Groves, C.R. *et al.* 2012. Incorporating climate change into systematic conservation planning. *Biodivers. Conserv.* **21:** 1651–1671.

106. Pressey, R.L. & M.C. Bottrill. 2008. Opportunism, threats, and the evolution of systematic conservation planning. *Conserv. Biol.* **22:** 1340–1345.

107. Barlow, C. 2012. Rewilding *Torreya Taxifolia.* www.Torreya Guardians.org.

108. Weible, C.M. *et al.* 2012. Understanding and influencing the policy process. *Policy Sci.* **45:** 1–21.

109. Weible, C.M., P.A. Sabatier & K. McQueen. 2009. Themes and variations: taking stock of the advocacy coalition framework. *Policy Studies J.* **37:** 121–140.

110. Leach, W.D. & P.A. Sabatier. 2005. To trust an adversary: integrating rational and psychological models of collaborative policymaking. *Am. Pol. Sci. Rev.* **99:** 491–503.

111. Lubell, M. 2007. Familiarity breeds trust: collective action in a policy domain. *J. Pol.* **69:** 237–250.

112. Sabatier, P.A. & L.K. Shaw. 2009. Are collaborative watershed management groups democratic? An analysis of California and Washington partnerships. *J. Soil Water Conserv.* **64:** 61A–64A.

113. Conant, S. 1988. Saving endangered species by translocation. *Bioscience* **38:** 254–257.

114. McLane, S.C. & S.N. Aitken. 2012. Whitebark pine (Pinus albicaulis) assisted migration potential: testing establishment north of the species range. *Ecol. Appl.* **22:** 142–153.

115. Roncal, J. *et al.* 2012. Testing appropriate habitat outside of historic range: the case of Amorpha herbacea var. crenulata (Fabaceae). *J. Nat. Conserv.* **20:** 109–116.

116. Olden, J.D. *et al.* 2006. The rapid spread of rusty crayfish (Orconectes rusticus) with observations on native crayfish declines in Wisconsin (USA) over the past 130 years. *Biol. Invas.* **8:** 1621–1628.

117. Drechsler, M. 2004. Model-based conservation decision aiding in the presence of goal conflicts and uncertainty. *Biodivers. Conserv.* **13:** 141–164.

118. McDonald-Madden, E., P.W. J. Baxter & H.P. Possingham. 2008. Making robust decisions for conservation with restricted money and knowledge. *J. Appl. Ecol.* **45:** 1630–1638.

119. McLean, I.F. G. 2003. A policy for conservation translocation of species in Britain. Joint Nature Conservation Committee on behalf of The Countryside Council for Wales, E.N. a. S.N. H., Ed.: 34. Peterborough.

120. Parker, K.A. *et al.* 2012. The theory and practice of catching, holding, moving and releasing animals. In *Reintroduction Biology: Integrating Science and Management.* J.G. Ewen *et al.* Eds.: Wiley Blackwell. Chichester, UK.

121. Jones, C.G. & D.V. Merton. 2012. A tale of two islands: the rescue and recovery of endemic birds in New Zealand and Mauritius. In *Reintroduction Biology: Integrating Science and Management.* J.G. Ewen *et al.,* Eds.: 33–72. Wiley Blackwell. Chichester, UK.

122. Bernardo, C.S. S. *et al.* 2011. Using post-release monitoring data to optimize avian reintroduction programs: a 2-year case study from the Brazilian Atlantic Rainforest. *Ani. Conserv.* **14:** 676–686.

123. Lewis, J.C., R.A. Powell & W.J. Zielinski. 2012. Carnivore translocations and conservation: insights from population models and field data for fishers (Martes pennanti). *Plos One* **7:** 1.

124. Nichols, J.D. & D.P. Armstrong. 2012. Monitoring for reintroductions. In *Reintroduction Biology: Integrating Science and Management.* J.G. Ewen *et al.,* Eds.: 223–256. Wiley Blackwell. Chichester UK.

125. Chauvenet, A.L.M. *et al.* 2012. Does supplemental feeding affect the viability of translocated populations? The example of the hihi. *Anim. Conserv.* **15:** 337–350.

126. Mawdsley, J. 2011. Design of conservation strategies for climate adaptation. *Wiley Interdiscipl. Rev.-Clim. Change* **2:** 498–515.

127. Jenkins, L.D., S.M. Maxwell & E. Fisher. 2012. Increasing conservation impact and policy relevance of research through embedded experiences. *Conserv. Biol.* **26:** 740–742.

128. Sutherland, W.J. *et al.* 2011. Quantifying the Impact and Relevance of Scientific Research. *Plos One.* **6:** e27537.

129. International Union for Conservation of Nature (IUCN). 2012. Guidelines for Reintroductions and Other Conservation Translocations. Species Survival Commission. http://www.issg.org/pdf/publications/Translocation-Guidelines-2012.pdf.

130. Boyce, M.S. 1992. Population viability analysis. *Annu. Rev. Ecol. Sys.* **23:** 481–506.

131. Gregory, A.J. *et al.* 2012. Influence of translocation strategy and mating system on the genetic structure of a newly established population of island ptarmigan. *Conserv. Genet.* **13:** 465–474.

132. Dunham, J. *et al.* 2011. Assessing the feasibility of native fish reintroductions: a framework applied to threatened bull trout. *North Am. J. Fisheries Manage.* **31:** 106–115.

133. Groce, M.C., L.L. Bailey & K.D. Fausch. 2012. Evaluating the success of Arkansas darter translocations in Colorado: an occupancy sampling approach. *Trans. Am. Fisheries Soc.* **141:** 825–840.

134. MacKenzie, D.I. *et al.* 2003. Estimating site occupancy, colonization, and local extinction when a species is detected imperfectly. *Ecology* **84:** 2200–2207.

135. La Morgia, V. *et al.* 2011. Where do we go from here? Dispersal simulations shed light on the role of landscape structure in determining animal redistribution after reintroduction. *Landscape Ecol.* **26:** 969–981.

Ann. N.Y. Acad. Sci. ISSN 0077-8923

ANNALS OF THE NEW YORK ACADEMY OF SCIENCES

Issue: *The Year in Ecology and Conservation Biology*

The Marine Mammal Protection Act at 40: status, recovery, and future of U.S. marine mammals

Joe Roman,[1] Irit Altman,[2] Meagan M. Dunphy-Daly,[3] Caitlin Campbell,[4] Michael Jasny,[5] and Andrew J. Read[3]

[1]Gund Institute for Ecological Economics, University of Vermont, Burlington, Vermont. [2]Biology Department, Boston University, Boston, Massachusetts. [3]Division of Marine Science and Conservation, Nicholas School of the Environment, Duke University, Beaufort, North Carolina. [4]Biology Department, University of Vermont, Burlington, Vermont. [5]Natural Resources Defense Council, Santa Monica, California

Address for correspondence: Joe Roman, Gund Institute for Ecological Economics, University of Vermont, Burlington, VT 05405. romanjoe@gmail.com

Passed in 1972, the Marine Mammal Protection Act has two fundamental objectives: to maintain U.S. marine mammal stocks at their optimum sustainable populations and to uphold their ecological role in the ocean. The current status of many marine mammal populations is considerably better than in 1972. Take reduction plans have been largely successful in reducing direct fisheries bycatch, although they have not been prepared for all at-risk stocks, and fisheries continue to place marine mammals as risk. Information on population trends is unknown for most (71%) stocks; more stocks with known trends are improving than declining: 19% increasing, 5% stable, and 5% decreasing. Challenges remain, however, and the act has generally been ineffective in treating indirect impacts, such as noise, disease, and prey depletion. Existing conservation measures have not protected large whales from fisheries interactions or ship strikes in the northwestern Atlantic. Despite these limitations, marine mammals within the U.S. Exclusive Economic Zone appear to be faring better than those outside, with fewer species in at-risk categories and more of least concern.

Keywords: Endangered Species Act; marine mammals; Marine Mammal Protection Act; status and trends; stock assessment reports

Introduction

Legislation protecting whales dates back to 1934, when right whale hunting was banned by an international treaty. In the early 1970s, further attempts to protect the great whales in the United States were met with resistance by the U.S. Department of Defense, which was concerned about the supply of sperm whale oil for use as a lubricant in submarines and other military engines. After a synthetic oil was produced, the Marine Mammal Protection Act (MMPA) was passed in October 1972. The MMPA went beyond protection for commercial reasons and attempted to restore the ecological role of all marine mammals. It was a critical step toward the passage of the Endangered Species Act (ESA) the following year.[1]

The fundamental objectives of the MMPA are (1) to maintain stocks of marine mammals at their optimum sustainable populations (OSP) and (2) to maintain marine mammal stocks as functioning elements of their ecosystems. The act does not define OSP, but the National Marine Fisheries Service (NMFS) has interpreted OSP to be a population level that falls between Maximum Net Productivity Level (MNPL) and carrying capacity (K). In operational terms, therefore, OSP is defined as a population size that falls between $0.5K$ and K. In addition, there is a clear mandate to protect individual marine mammals from harm, referred to as *take*.

In this review, we assess the success of the MMPA in protecting marine mammals, discuss its failures, and provide suggestions on ways to improve the act and marine mammal conservation in the United States and internationally.

By the numbers

U.S. marine mammal stocks 1995–2011

In the United States, two federal agencies direct the management and protection of marine mammals:

doi: 10.1111/nyas.12040

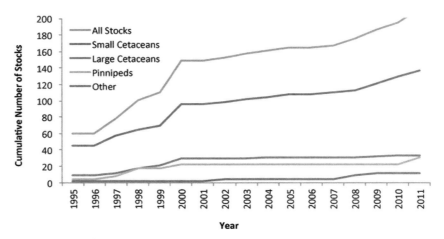

Figure 1. Cumulative number of stocks recognized under the Marine Mammal Protection Act since stock assessment reports began in 1995.

NMFS is responsible for managing most marine mammal stocks, including cetaceans, sea lions, and seals; the U.S. Fish and Wildlife Service (USFWS) has authority over a smaller number of stocks that include polar bears, sea otters, manatees, and walruses. Under the MMPA, a marine mammal stock is defined as a group of individuals "of the same species or smaller taxa in a common spatial arrangement that interbreed when mature." Stock assessment reports (SARs) for all marine mammals that occur in U.S. waters were first required when the act was amended in 1994. Since that time, all stocks have been reviewed at least every three years or as new information becomes available. Stocks that are designated as *strategic* are reviewed annually. Each draft SAR is peer-reviewed by one of three regional Scientific Review Groups (SRGs) and revised and published after a public comment period. These reports are extremely valuable for the information they provide and their transparency: documents are posted online (www.nmfs.noaa.gov/pr/sars/). During the 17 years that the agencies have conducted SARs, many new stocks have been recognized (Fig. 1), and information about the demography and distribution of existing populations has led to many cases of stock reclassification. In some cases, reclassified stocks leave older stocks obsolete; for example, if a single, large stock is recognized to be composed of multiple, small, and discrete breeding populations. In other cases, a remnant of the original stock may still be considered, even while a subset of the population is designated as an independent stock.

We examined the history of marine mammal stock classification over time (1995–2011), taking into account newly recognized stocks and the dissolution of older stocks. A cumulative frequency analysis shows that the number of recognized stocks for all groups of marine mammals increased rapidly in the early years of assessment, with most stocks designated between 1995 and 2000 (Fig. 1). After 2000, few additional stocks were identified for species of large cetaceans. The number of pinniped stocks increased as a result of the reclassification of Alaska harbor seal stocks from 3 to 12 distinct stocks in 2011. Similarly, USFWS stocks exhibited a slight increase in the number of recognized stocks in recent years owing mainly to the classification of sea otter. In contrast to the relatively small changes in stock classification of these groups since 2000, the number of small cetacean stocks exhibited a large increase. Since 2000, the annual rate of increase in the number of newly identified small cetacean stocks has been more than four times that of other groups (small cetaceans = 3.7; large cetaceans = 0.4; USFWS species = 0.8; pinnipeds = 0.8 newly designated stocks per year), suggesting that either information on the population structure of small cetaceans is increasing faster than for other taxa or that odontocetes have finer population structure than other marine mammals.

In 2011, a total of 212 stocks of marine mammals were designated under the management authority of NMFS and USFWS, of which most (65%) were small cetaceans. Large cetaceans represent the second largest group, accounting for nearly 16% of all stocks. Pinnipeds account for 15% of all current stocks, with the remainder being species managed by USFWS.

Population trends

For all currently recognized marine mammal stocks, we reviewed the earliest and most recent stock assessments to investigate trends in abundance. For many stocks, information on abundance is limited and even less is known about trends. It should be noted, however, that identifying trends in marine mammals is known to be difficult. Taylor *et al.*, for example, found that even precipitous declines would not be noticed for 72% of large whale stocks, 78% of dolphins and porpoises, and all pinnipeds counted on ice with current levels of survey effort.[2] Declines in land-based pinnipeds were much easier to detect. Whereas the MMPA does not require information on trends, stock assessment reports can describe a variety of available information on abundance trends, including information from the literature, unpublished data, and expert insight. Pulling all this information into a single document is useful and important, but obtaining a formal assessment of trends over time is often restricted by inconsistencies in the methods of multiple independent studies and limited understanding of patterns across the whole spatial range of the stock. Despite these limitations, it is important to analyze the evidence available on marine mammal trends since this is an essential metric for assessing the health of these populations.

For this work, we examined descriptions in the earliest and latest SARs and classified the presence and direction of the most recent trends identified for a stock. We noted if the trend was definitively stated in the SAR or if the description indicated a possible trend. We summarized trends with respect to the following categories: decreasing, increasing (including cases where the stock is classified as "stable or increasing"), stable, and unknown.

Information on population trends is currently unknown for the majority (71%) of U.S. marine mammal stocks. Ten percent of stocks currently exhibit increasing abundance trends with the percentage increasing to 19% when possible cases of increases are included. Two percent of stocks currently exhibit stable trends, which increases to 5% when stocks with possible stable trends are included. Three percent of stocks exhibit decline,

which increases to 5% if possible declining trends are included (see Supporting Information).

Overall, the pattern is consistent with trends from the earliest years in which stocks were assessed. In the first year that each stock was assessed, 68% of the stocks had unknown trends; 7% showed evidence of increase, increasing to 21% when possible trends are included; 2% were stable, increasing to 5% when possible trends are included; and 4% were found to be decreasing, increasing to 6% when possible trends are included (see Supporting Information).

For stocks that exhibited a definitive trend in the earliest year in which they were assessed, we examined whether the population continued to show a similar trend in the most recent SAR. The majority of stocks in this group exhibit no change in the direction of the trend between the earliest and latest SAR. A total of seven stocks were found to exhibit stable or increasing trends in the earliest and latest years in which they have been assessed. Two stocks were described as declining in the earliest and latest SAR. Two other stocks demonstrated reversal of trends in the earliest and latest SARs. The Oregon–Washington coastal harbor seal was described as decreasing in the earliest SAR, then stable or increasing in abundance in the most recent report. In contrast, the Eastern Pacific northern fur seal was identified as stable in the earliest SAR and decreasing in the most recent report.

Status and trends: endangered species

We examined the status under the ESA for all current stocks of marine mammals using information from the latest SAR and additional sources of information (www.nmfs.noaa.gov/pr/species/esa/other.htm). Of the 212 current stocks, 38 (18%) are listed as endangered or threatened (Supporting Information Appendix 1). Although small cetaceans have the greatest number of stocks in U.S. waters, only two of them are listed under the ESA: the Southern resident eastern North Pacific killer whale and the recently listed Hawaiian insular false killer whale. In contrast, 25 stocks of large cetaceans (representing 76% of this group) are currently listed as threatened or endangered (Fig. 2A). Of the two other groups, four pinniped stocks (13%) and seven USFWS managed stocks (64%) are listed under the ESA.

The highest number and proportion of threatened or endangered stocks are found in the Pacific region, where 21 of the 81 stocks are listed (26%).

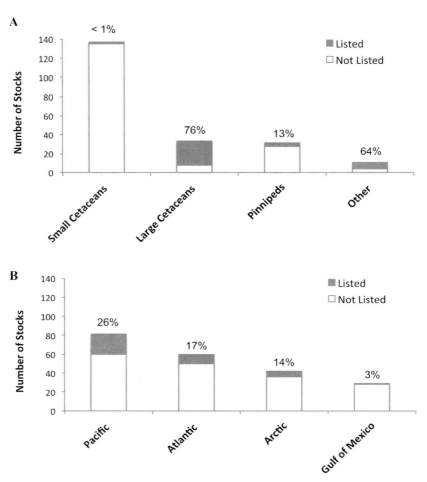

Figure 2. Number of marine mammal stocks protected by the Endangered Species Act, (A) by taxonomic group, (B) by geographical region. The percentage of ESA-protected stocks for each group is presented above the bars.

In the Atlantic region, 10 of 60 stocks are ESA listed, representing 17% of stocks found in this region. In the Arctic region six of 42 stocks (14%) are ESA listed. In the Gulf of Mexico, only the Northern Gulf of Mexico sperm whale is listed under the ESA (Fig. 2B).

To determine how the ESA status of marine mammals changed over time, we compared information from the earliest and most recent year each stock was assessed. The majority of the 38 stocks currently listed under the ESA were also listed at the time of their first assessment. Three stocks, however, became listed only in the most recent years in which they were assessed: the Alaska Chukchi/Bering Seas polar bear stock (which was designated as a stock in 2002 and became listed as threatened in 2008), the eastern North Pacific Southern resident false

killer whale (designated as a stock in 1999, listed as endangered in 2005), and the Hawaiian insular false killer whale, which was classified as endangered in 2012. Endangered species listing is pending for three stocks. Recent petitions include the Alaskan Pacific walrus (2009), two distinct population segments of bearded seals associated with the Alaska stock (2010), and four subspecies of ringed seal associated with the Alaska stock (2010).

We did not identify any case in which a stock was listed as threatened or endangered in the earliest stock assessment and then delisted in the most recent assessment, but the Eastern stock of Steller sea lion was proposed for delisting in 2012 because the stock is thought to have recovered and the Gulf of Maine–Bay of Fundy harbor porpoise was proposed as a threatened species in 1995 and removed as a candidate by 2011. One marine mammal stock recovered before its first assessment—the eastern North Pacific gray whale, which was delisted in 1994. In addition, the Caribbean monk seal was delisted because the species was formally recognized as extinct. Efforts to investigate unconfirmed sightings of Caribbean monk seals since the species was first listed under the Endangered Species Protection Act in 1967 (and relisted under the ESA in 1979) revealed only extralimital northern seals. Last seen in 1952, the Caribbean monk seal was almost certainly extinct at the time of passage of the MMPA and ESA. Officially delisted in 2008,[3] it is the only known case of a recent extinction for a U.S. marine mammal.

Status and trends: strategic stocks that exceed potential biological removal

Potential biological removal (PBR) is the critical threshold defined under the MMPA as the maximum number of animals, not including natural mortalities, that may be removed from a marine mammal stock while allowing that stock to recover to or be maintained within its OSP. PBR is defined as the product of the minimum population estimate ($N_{\min}$), half the maximum net productivity rate ($R_{\max}$), and a recovery factor (F_r), which ranges from 0.1 to 1.0.[4] We examined stocks for which human-associated mortality exceeded PBR (or was very likely to exceed PBR) in either the earliest or the most recent year in which PBR was determined in a SAR. Occasionally information necessary to determine PBR was not available in the earliest or most recent SAR; in such cases, we examined all relevant SARs to find the earliest and latest years in which PBR was reported. For stocks that exceed PBR, we obtained the best available estimates of human-related mortality provided in the SAR and listed the primary sources of mortality.

We found nine improved stocks, for which mortality exceeded PBR in the original assessment but was less than PBR in the most recent SAR (Table 1A). Sixteen stocks are currently exceeding PBR based on the most recent information available (Table 1B–D). Of these, 9 (56%) show no change in status with respect to PBR. The most common sources of mortality for these stocks are fishing interactions and ship strikes (Table 1B). Four stocks have degraded, exceeding PBR in the most recent assessment but not the earliest for which information was available. The primary mortality sources

for this group are also fisheries interactions (gillnets) and ship strikes (Table 1C). The remaining two stocks exceeding PBR are recently designated and have only been assessed once (Table 1D). Finally, there are three stocks in which mortality exceeded PBR in the earliest SAR, but no designation was made in the latest assessment because of insufficient mortality information (Table 1E).

Strategic stocks

Stocks that are listed under ESA and those where human-related mortality exceeds PBR are automatically considered *strategic* by NMFS and USFWS. In addition, a stock may be considered strategic if there is evidence that the population is declining and likely to be listed under the ESA in the foreseeable future.

A total of 76 stocks (i.e., 36% of all recognized stocks) are currently identified as strategic, including cases where stocks are identified as *probably strategic*, as with the false killer whale stock in American Samoa. Human mortality exceeds PBR for 15 of these stocks, based on the most recent information available (Table 1B–D). Thirty-five strategic stocks are considered depleted under MMPA, even though anthropogenic mortality is not currently known to exceed PBR. This category predominantly includes small cetaceans, such as bottlenose dolphins, and pinnipeds, such as Alaskan harbor seals, many of which are recently designated stocks for which there may be limited information to determine PBR and mortality. In these cases, strategic designation provides an added layer of protection when definitive data on population metrics do not exist. The remaining 26 strategic stocks are not considered depleted, nor are they known to experience human-related mortality exceeding PBR. For these cases, limited information and small population size may warrant classifying the stock as strategic until clear evidence can be gathered that it is not at risk. Some stocks with limited data and unknown population size, however, are not classified as strategic if it appears that abundance is high and human-related mortality is low.

Status of U.S. marine mammals species: a global perspective

To assess the relative success of marine mammal protection in the United States under the MMPA and ESA, we compared the status of marine mammal species found within the U.S. Exclusive Economic

Table 1. U.S. marine mammal stocks for which human-influenced mortality exceeds (or is very likely to exceed) potential biological removal (PBR), either in the latest or earliest year in which information is available from the stock assessment report (SAR). Stocks that demonstrate improvement over time, where human-influenced mortality exceeded PBR in the earliest but not the most recent SAR are shown in group A. Group B includes stocks that exhibit no change with respect to exceeding PBR in the earliest and latest SAR. Stocks that exceed PBR only in the most recent SAR but not in the earliest SAR are shown in C. Recently designated stocks (with only a single assessment) for which mortality exceeds PBR are shown in D. Group E represents stocks for which PBR is exceeded in the earliest year and no designation can be made with respect to exceeding PBR in the most recent year. Values for PBR and mortality from relevant SARs are presented. Primary sources of mortality are also listed when this information is provided in the SAR.

Group	Region	Species	Stock	Earliest year of SAR w/PBR	Earliest PBR/ mortality	Latest year of SAR w/PBR	Latest PBR/ mortality	Primary mortality sources
A								
	Atlantic	Common dolphin, short-beaked	Western North Atlantic	1995	32/449	2011	1,000/164	Fishing interactions (gillnet),[e] ship strikes,[e] whaling historic[e]
		Spotted dolphin, Pantropical	Western North Atlantic	1995	UNK/31	2007	30/7	Fishing interactions (gillnet),[e] ship strikes[e]
		Pilot whale, short-finned	Western North Atlantic	1995	3.7/UNK	2011	93/UNK	Fishing interactions (gillnet),[e] strandings[e]
	Pacific	Humpback whale[*d]	California– Oregon– Washington	1999	0.8/2	2010	11.3/3.6	
		Sperm whale[*d]	California– Oregon– Washington	1999	2/3	2010	1.5/0.4	Fishing interactions (gillnet)[e]
		Harbor porpoise	Monterey Bay	2002	11/80	2009	10/UNK	Fishing interactions (gillnet)[e]
		Pilot whale, short-finned	California– Oregon– Washington	1999	6.9/13	2010	4.6/0	Subsistence fishing[e]
		Steller sea lion[*d]	Western	1998	350/444	2011	253/227.1	Fishing interactions (gillnet, squid)[e]
	Arctic	Beluga whale	Cook Inlet	1998	14/72	2005	2/0	Fishing interactions (gillnet, trawl),[e] subsistence fishing[e]
B								
	Atlantic	Right whale, North Atlantic[*d]	Western Stock	1995	0.4/2.6	2011	0.8/2.4	Fishing interactions (gillnet),[e] ship strikes,[l] fishing interactions (unidentified)[l]
		Sei whale[*d]	Nova Scotia	2007	0.3/0.4	2011	0.4/1.2	Fishing interactions (gillnet, unidentified)[e,l]

Continued

Table 1. *Continued*

Group	Region	Species	Stock	Earliest year of SAR w/PBR	Earliest PBR/ mortality	Latest year of SAR w/PBR	Latest PBR/ mortality	Primary mortality sources
		Harbor porpoise[$]	Gulf of Maine-Bay of Fundy	1995	403/1,876	2011	701/927	Pollution,[e] fishing interactions (gillnet),[e,l] strandings,[e,l] fishing interactions (trawl, mackerel)[l]
		White-sided dolphin, Atlantic	Western North Atlantic	1995	125/127	2011	190/245	Fishing interactions (gillnet, longline, unidentified)[e,l]
		West Indian manatee[*d]	Antillean	1995	0/5	2009	0.144/8.2	Fishing interactions (longline),[e,l] boat strikes[l]
		West Indian manatee[*d]	Florida	1995	0/40.1	2009	11.8/86.6	Fishing interactions (longline),[e,l] boat strikes[l]
	Pacific	False killer whale	Pacific Islands Region Stock Complex - Hawaii	2000	0.8/9	2007	2.4/4.9	Fishing interactions (unidentified),[e,l] ship strikes[e,l]
		False killer whale	Pacific Islands Region Stock Complex - Hawaii Pelagic	2008	2.2/5.7	2011	2.4/10.8	Ship strikes[e,l]
	Arctic	Pacific walrus[¥]	Alaska	2009	2,580/5,460	2010	2,580/ 5,457	Fishing interactions (trawl, unidentified),[e,l] habitat/oil/gas[e,l]
C	*Atlantic*	Humpback whale[*d]	Gulf of Maine	1995	9.7/1	2011	1.1/5.2	Ship strikes[l]
	Gulf of Mexico	Bryde's whale	Northern Gulf of Mexico	1995	0.2/UNK	2011	0.1/1	Fishing interactions (gillnet),[e,l] ship strikes,[e,l] toxins from harmful algal blooms[e,l]
	Pacific	False killer whale[*]	Pacific Islands Region Stock Complex - Hawaii Insular	2008	0.8/0	2011	0.2/0.6	Fishing interactions (gillnet)[l]
		Killer whale[*d]	Eastern North Pacific Southern Resident	1999	0.9/0	2011	0.17/0.2	None reported[l]

Continued

Table 1. *Continued*

Group	Region	Species	Stock	Earliest year of SAR w/PBR	Earliest PBR/ mortality	Latest year of SAR w/PBR	Latest PBR/ mortality	Primary mortality sources
D								
	Atlantic	West Indian manatee[*d]	Puerto Rico	2009	0.144/8.2	na	na	Fishing interactions (gillnet, squid, mackerel),[e] fishing interactions (longline, trawl, groundfish)[l]
	Pacific	False killer whale	American Samoa	2010	7.5/7.8	na	na	Fishing interactions (gillnet),[e] ship strikes[e]
E								
	Atlantic	Bottlenose dolphin	W.N. Atlantic Offshore	1995	92/128	2008	566/UNK	Fishing interactions (gillnet, trawl)[e]
		Bottlenose dolphin[d]	W.N. Atlantic Northern Migratory Coastal	2002	23/30	2010	71/UNK	
		Pilot whale, long-finned	Western North Atlantic	1995	28/UNK	2011	93/UNK	Pollution,[e] Fishing interactions (gillnet),[e,l] strandings,[e,l] fishing interactions (trawl)[l]

NOTES: We followed designations within the SAR for exceeding potential biological removal (PBR) and indicated cases when mortality or PBR is unknown (UNK). For these cases, information from previous assessments is sometimes used to suggest whether PBR is likely exceeded. In addition, there are cases where small population size (and thus presumably small PBR) warrants a designation of PBR exceeded. Symbols after species names indicate that the stock is currently listed under ESA (*); petitioned for listing under the ESA (¥); recently removed as a candidate for ESA listing ($); and currently depleted under the MMPA (d). Letters after primary sources of mortality indicate (e) the mortality source was relevant in the earliest SAR and (l) the mortality source was relevant in the latest SAR.

Zone (EEZ) to those outside of the U.S. EEZ, using the most recent designations (1996–2012) provided by the International Union for Conservation of Nature (IUCN). The total number of marine mammal species associated with the two groups was nearly equal (number of U.S. marine mammal species = 65, non-U.S. species = 67) and results indicate that U.S. species generally fare better than non-U.S. species in all categories (Fig. 3). Specifically, fewer U.S. species are found in high-risk categories (vulnerable, critically endangered, near threatened, extinct) and more U.S. species are considered of least concern. In such an uncontrolled comparison, it is impossible to draw definitive conclusions regarding the factors responsible for this difference; nevertheless, the patterns suggest fundamental prohibitions against the taking of marine mammals in the MMPA, along with the ESA, likely contribute to this difference. We conclude that marine mammals found in the United States do appear to be doing as well and in many cases better than species found outside of U.S. waters, suggesting that current management actions are having a positive influence on marine mammal populations.

sea-ice habitat, which could lead to higher density of hosts and favor density-dependent disease; decreases in food availability leading to impacts on body condition and immune system function; and increases in human activity throughout the region leading to increased likelihood of pathogen introduction. Finally, increased susceptibility to disease has recently been linked to decreased genetic variability in populations of California sea lions,[94] underscoring the heightened threat to endangered populations. Anthropogenically exacerbated diseases in pinnipeds, cetaceans, and sea otters, from harmful algal blooms to pathogen pollution from pets and livestock, demonstrate that the protection of marine mammals also requires protection of the adjacent terrestrial environment.

The way forward

The MMPA has been very successful in protecting many marine mammals from harm and largely successful in restoring and protecting individual marine mammals stocks. One of the reasons for this success has been the development of the PBR approach by NMFS, designed expressly for management under the act. This current focus on species and individual animals is appropriate not only from a welfare perspective but also, given the lack of data and the need for precaution, from a demographic standpoint.

There have been few, if any, attempts to address the second fundamental objective: maintaining marine mammals as functional elements of their ecosystem. Many species lack historic baselines, and the understanding of the ecological role of marine mammals was limited when the act was passed. It is increasingly clear, however, that upper trophic level predators, such as marine mammals, play critical roles in structuring their ecosystems.[95–97] Humpback and fin whales in the Gulf of Maine increase productivity by pumping nutrients to the surface.[6] The benthic plowing of gray whales alters the microtopography of the seafloor and enhances benthic-pelagic coupling.[98] Estes *et al.* have even suggested that productive and dense kelp forests can be used as a sensitive and cost-effective measure of sea otter recovery, an approach that has broad potential in establishing recovery criteria for other reduced populations with clearly measurable ecosystem impacts.[99]

To restore the ecological role of marine mammals, there is a need for an ecoregional approach to conservation, with an increased understanding of predator–prey interactions and the cumulative effects of human impacts. A precautionary generalization of PBR that combines the direct and indirect effects of fisheries, including predator–prey relationships and ecological interactions, as well as cumulative impacts from other stressors, could form a central part of such policy. Such an effort would balance the apparently competing management goals of optimum fishery yield and sustainable marine mammal populations. This would, of course, require a fundamental rethinking of how we manage fisheries and other extractive and nonextractive ocean uses.

Our increased understanding of the stock structure of marine mammal populations has clearly aided in our ability to manage them. The number of U.S. stocks has more than tripled since SARs were first compiled, largely because of a better comprehension of odontocete population structure (Fig. 1). Assessing the status of marine ecoregions together with the dynamics of these individual stocks would represent a significant step forward in ocean conservation. Such a comprehensive management framework would move the species-based approach to one that can effectively restore the ecological function of marine mammals. Whales and other long-lived species can dampen the frequency and amplitude of oscillations from perturbations in climate, predation, and primary productivity.[97] The removal of these species from much of the world has left many marine communities dominated by *r*-selected species. Without whales, marine ecosystems have longer return time after perturbations.

The MMPA, along with the Endangered Species Act, has helped put several great whale species, including the Pacific gray whale, Pacific blue whale, and humpbacks in the Atlantic and Pacific, on the road to recovery, a process that was aided by the moratorium on commercial whaling by the great majority of nations. The restoration of whales and other marine mammals has been a great benefit to coastal communities in the United States, bringing more than $956 million a year in the form of whale watching,[100] increasing the diversity of jobs in areas suffering from fisheries decline, such as Gloucester and Provincetown, Massachusetts, and enhancing environmental tourism. The increase in whale watching has come at a cost, including collisions between whale-watching boats and whales and reduced reproductive fitness.[101,102] Other threats have also emerged or been acknowledged in the 40 years since the act was passed, including the rise of disease, ship collisions, declines in prey species, and noise and disturbance. Research and new technologies are clearly needed to protect marine mammals from noise-related impacts, including the study of behavioral responses to impulsive and continuous noise.[65]

The MMPA has focused on addressing direct effects, but it should be kept in mind that there are indirect consequences of restoration: you cannot have healthy marine mammal populations without a healthy marine ecosystem. In this way, a fully enforced MMPA could serve as a de facto marine conservation act, much as the ESA has become a habitat protection act, at least in terrestrial ecosystems. The restoration of marine mammals may go well beyond such legislative boundaries: as active members in the marine food web, they can help restore coastal and pelagic ecosystems simply by becoming functional members of marine communities.

Acknowledgments

We thank Brad Sewell for prompting a study of the status of U.S. marine mammals, and Rick Ostfeld and Bill Schlesinger, of the Cary Institute for Ecosystem Studies, for inviting us to prepare this review. Research was supported by the Natural Resources Defense Council's Endangered Oceans Project.

Supporting Information

Additional Supporting Information may be found in the online version of this article.
Appendix 1: Status and trends of U.S. marine mammal stocks

Conflicts of interest

The authors declare no conflicts of interest.

References

1. Roman, J. 2011. *Listed: Dispatches from America's Endangered Species Act.* Harvard University Press. Cambridge, MA.
2. Taylor, B.L. *et al.* 2007. Lessons from monitoring trends in abundance of marine mammals. *Mar. Mammal Sci.* **23:** 157–175.
3. NOAA. 2008. NOAA confirms Caribbean monk seal extinct. Available at http://www.noaanews.noaa.gov.
4. Wade, P.R. & R. Angliss. 1997. Guidelines for Assessing Marine Mammal Stocks: *Report of the GAMMS Workshop April 3–5, 1996, Seattle, Washington.* NOAA Technical Memorandum. NMFS-OPR-12.
5. Morissette, L., V. Christensen & D. Pauly. 2012. Marine mammal impacts in exploited ecosystems: would large scale culling benefit fisheries? *PLoS ONE* **7:** e43966.
6. Roman, J. & J.J. McCarthy. 2010. The whale pump: marine mammals enhance primary productivity in a coastal basin. *PLoS ONE* **5:** e13255.
7. Albertson, G.R. *et al.* 2011. Staying close to home? Genetic analyses reveal insular population structure for the pelagic dolphin *Steno bredanensis.* In *Abstracts of the 19th Biennial Conference on the Biology of Marine Mammals.* Tampa, Florida, USA.
8. Baird, R.W. *et al.* 2008. Site fidelity and association patterns in a deep-water dolphin: rough-toothed dolphins (Steno bredanensis) in the Hawaiian Archipelago. *Mar. Mammal Sci.* **23:** 535–553.
9. Schweder, T. *et al.* 2010. Population estimates from aerial photographic surveys of naturally and variably marked bowhead whales. *J. Agric. Biol. Envir. S.* **15:** 1–19.
10. Marine Mammal Commission. 2008. Annual report to Congress 2007. Available online at: http://www.mmc.gov. Accessed on Jan. 10, 2013.
11. Bodkin, J.L. & B.E. Ballachey. 2010. Modeling the effects of mortality on sea otter populations. U.S. Department of the Interior and U.S. Geological Survey. Scientific Investigations Report 2010–5096.
12. Read, A.J. 2008. The looming crisis: Interactions between marine mammals and fisheries. *J. Mammal* **89:** 541–548.
13. Carretta, J.V., J. Barlow & L. Enriquez. 2008. Acoustic pingers eliminate beaked whale bycatch in a gill net fishery. *Mar. Mammal Sci.* **24:** 956–961.
14. Davidson, A.D. *et al.* 2012. Drivers and hotspots of extinction risk in marine mammals. *Proc. Natl. Acad. Sci. U. S. A.* **109:** 3395–3400.
15. Orphanides, C.D. 2012. New England harbor porpoise bycatch rates during 2010–2012 associated with Consequence Closure Areas. US Department of Commerce, NEFSC Ref Doc. 12–19.
16. U.S. Government Accountability Office. 2008. National Marine Fisheries Service: improvements are needed in the federal process used to protect marine mammals from commercial fishing. Report no. GAO-09-78.
17. Geijer, C.K.A. & A.J. Read. 2013. Mitigation of marine mammal bycatch in U.S. Fisheries since 1994. *Biol. Conserv.* In press.
18. Marine Mammal Commission. 2007. The biological viability of the most endangered marine mammals and the cost-effectiveness of protection programs: a report to Congress by the Marine Mammal Commission. Available at: http://www.mmc.gov. Accessed on Jan. 10, 2013.
19. Gerber, L.R. *et al.* 2011. Managing for extinction? Conflicting conservation objectives in a large marine reserve. *Cons. Lett.* **4:** 417–422.
20. Brownell, R.L. *et al.* 2001. Conservation status of North Pacific right whales (*Eubalaena japonica*). *Cetacean Res. Manage.* **2:** 269–286.

21. Kennedy, A.S., D.R. Salden & P.J. Clapham. 2011. First high- to low-latitude match of an eastern North Pacific right whale (*Eubalaena japonica*). *Mar. Mammal Sci.* **28:** E539–E544.

22. Wade, P.R. *et al.* 2011. The world's smallest whale population? *Biol. Lett.* **7:** 83–85.

23. Knowlton, A.R. *et al.* 2012. Monitoring North Atlantic right whale *Eubalaena glacialis* entanglement rates: a 30 yr retrospective. *Mar. Ecol. Prog. Ser.* **466:** 293–302.

24. Van der Hoop, J.M. *et al.* 2012. Assessment of management to mitigate anthropogenic effects on large whales. *Conserv. Biol.* **27:** 121–133.

25. NOAA. 2012. Shifting the Boston traffic separation scheme. Available at: stellwagen.noaa.gov/science/tss.html. Accessed on Jan. 10, 2013.

26. Wiley, D.N. *et al.* 2011. Modeling speed restrictions to mitigate lethal collisions between ships and whales in the Stellwagen Bank National Marine Sanctuary, USA. **144:** 2377–2381.

27. Cox, T.M. *et al.* 2007. Comparing effectiveness of experimental and implemented bycatch reduction measures: the ideal and the real. *Conserv. Biol.* **21:** 1155–1164.

28. Cassoff, R.M. *et al.* 2011. Lethal entanglement in baleen whales. *Dis. Aquat. Org.* **96:** 175–185.

29. Marsh, H., J. O'Shea & J.E. Reynolds III. 2012. *Ecology and Conservation of the Sirenia: Dugongs and Manatees.* Cambrige University Press. Cambridge, UK.

30. Brosi, B.J. & E.G.N. Biber. 2012. Citizen involvement in the U.S. Endangered Species Act. *Science* **337:** 802–803.

31. Reeves, R.R., S. Leatherwood & R.W. Baird. 2009. Evidence of a possible decline since 1989 in false killer whales (*Pseudorca crassidens*) around the Main Hawaiian Islands. *Pac. Sci.* **63:** 253–261.

32. Robards, M.D. *et al.* 2009. Limitations of an optimum sustainable population or potential biological removal approach for conserving marine mammals: Pacific walrus case study. *J. Environ. Manage.* **91:** 57–66.

33. Aguilar, A. & A. Borrell. 1994. Abnormally high polychlorinated biphenyl levels in striped dolphins (*Stenella coeruleoalba*) affected by the 1990–1992 Mediterranean epizootic. *Sci. Total Environ.* **154:** 237–247.

34. Hall, A.J. *et al.* 1992. Organochlorine levels in common seals (*Phoca vitulina*) which were victims and survivors of the 1988 phocine distemper epizootic. *Sci. Total Environ.* **115:** 145–162.

35. Jepson, P.D. *et al.* 1999. Investigating potential associations between chronic exposure to polychlorinated biphenyls and infectious disease mortality in harbour porpoises from England and Wales. *Sci. Total Environ.* **244:** 339–348.

36. Jepson, P.D. *et al.* 2005. Relationships between polychlorinated biphenyls and health status in harbor porpoises (*Phocoena phocoena*) stranded in the United Kingdom. *Environ. Toxicol. Chem.* **24:** 238–248.

37. Ross, P.S. 2002. The role of immunotoxic environmental contaminants in facilitating the emergence of infectious diseases in marine mammals. *Human Ecol. Risk Assess.* **8:** 277–292.

38. Van Bressem, M.F. *et al.* 2009. Emerging infectious diseases in cetaceans worldwide and the possible role of environmental stressors. *Dis. Aquat. Org.* **86:** 143–157.

39. Krahn, M.M. *et al.* 2009. Effects of age, sex and reproductive status on persistent organic pollutant concentrations in "Southern Resident" killer whales. *Mar. Pollut. Bull.* **58:** 1522–1529.

40. Fair, P.A. *et al.* 2010. Contaminant blubber burdens in Atlantic bottlenose dolphins (Tursiops truncatus) from two southeastern US estuarine areas: concentrations and patterns of PCBs, pesticides, PBDEs, PFCs, and PAHs. *Sci. Total Environ.* **408:** 1577–1597.

41. Kannan, K. *et al.* 2004. Organochlorine pesticides and polychlorinated biphenyls in California sea lions. *Environ. Pollut.* **131:** 425–434.

42. Ylitalo, G.M. 2005. The role of organochlorines in cancer-associated mortality in California sea lions (Zalophus californianus). *Mar. Pollut. Bull.* **50:** 30–39.

43. Becker, P.R. *et al.* 1997. Concentrations of chlorinated hydrocarbons and trace elements in marine mammal tissues archived in the US National Biomonitoring Specimen Bank. *Chemosphere.* **34:** 2067–2098.

44. Reddy, M.L. *et al.* 2001. Opportunities for using Navy marine mammals to explore associations between organochlorine contaminants and unfavorable effects on reproduction. *Sci. Total Environ.* **274:** 171–182.

45. Morissette, L., K. Kaschner & L.R. Gerber. 2010. Ecosystem models clarify the trophic role of whales in Northwest Africa. *Mar. Ecol. Prog. Ser.* **404:** 289–303.

46. Estes, J.A. & G.R. VanBlaricom. 1985. Sea otters and shellfisheries. In *Marine Mammals and Fisheries.* J.R. Beddington, R.J.H. Beverton & D.M. Lavigne, Eds.: 187–235. George Allen & Unwin Ltd. London.

47. Moore, J. 2012. Management reference points to account for direct and indirect impacts of fishing on marine mammals. *Mar. Mammal Sci.* doi: 10.1111/j.1748-7692.2012.00586.x.

48. Hutchings, J.A. 2000. Collapse and recovery of marine fishes. *Nature* **406:** 882–885.

49. Christensen, V. *et al.* 2003. Hundred-year decline of North Atlantic predatory fishes. *Fish Fisheries* **4:** 1–24.

50. Myers, R.A. & B. Worm. 2003. Rapid worldwide depletion of predatory fish communities. *Nature* **423:** 280–283.

51. Plagányi, E.E. & D.S. Butterworth. 2005. Indirect fishery interactions. In *Marine Mammal Research: Conservation Beyond Crisis.* J.E. Reynolds, W.F. Perrin, R.R. Reeves, S. Montgomery & T.J. Ragen, Eds.: 19–48. The Johns Hopkins University Press. Baltimore, MA, USA.

52. Atkinson, S., D.P. DeMaster & D.G. Calkins. 2008. Anthropogenic causes of the western Steller sea lion *Eumetopias jubatus* population decline and their threat to recovery. *Mammal Rev.* **38:** 1–18.

53. Frank, K.T. *et al.* 2005. Trophic cascades in a formerly cod-dominated ecosystem. *Science* **308:** 1621–1623.

54. Cooper, F.-M. 2008. Settlement agreement. Filed in NRDC v. Winter, Case No. 05-cv-07513-FMC (C.D. Cal.).

55. NMFS. 2009. U.S. Navy's Atlantic Fleet Active Sonar Training (AFAST) Final Rule. *Federal Register* **74:** 4844–4885.

56. NMFS. 2012. Incidental take authorizations. Available at: http://www.nmfs.noaa.gov/pr/permits/incidental.htm. Accessed on Jan. 10, 2013.

57. McCarthy, E. 2004. *International Regulation of Underwater Sound: Establishing Rules and Standards to Address Ocean*

Noise Pollution. Kluwer Academic Publishers. Norwell, MA.

58. Richardson, W.J. *et al.* 1995. *Marine Mammals and Noise.* Academic Press. New York.

59. NMFS & U. S. Navy. 2001. Joint Interim Report: *Bahamas Marine Mammal Stranding Event of 15–16 March 2000.* Dept. of Commerce and U.S. Navy.

60. Fernández, A. *et al.* 2005. 'Gas and fat embolic syndrome' involving a mass stranding of beaked whales (family Ziphiidae) exposed to anthropogenic sonar signals. *Vet. Pathol.* **42**: 446–457.

61. Weilgart, L.S. 2007. The impacts of anthropogenic ocean noise on cetaceans and implications for management. *Can. J. Zool.* **85**: 1091–1116.

62. Wright, A.J. *et al.* 2007. Do marine mammals experience stress related to anthropogenic noise? *Int. J. Comp. Psychol.* **20**: 274–316.

63. Convention on Biological Diversity. 2012. Scientific synthesis on the impacts of underwater noise on marine and coastal biodiversity and habitats. UNEP/CBD/SBSTTA/16/INF/12. Montreal, Quebec, Canada.

64. Hatch, L.T. *et al.* 2012. Quantifying loss of acoustic communication space for right whales in and around a U.S. National Marine Sanctuary. *Conserv. Biol.* **26**: 983–994.

65. Daly, J.N. & J. Harrison. 2012. The Marine Mammal Protection Act: a regulatory approach to identifying and minimizing acoustic-related impacts on marine mammals. In *The Effects of Noise on Aquatic Life.* A.N. Popper & A. Hawkins, Eds.: 537–540.

66. Lusseau, D., L. Slooten & R.J.C. Currey. 2006. Unsustainable dolphin-watching tourism in Fiordland, New Zealand. *Tourism Mar. Enviro.* **3**: 173–178.

67. Clark, C.W. & G.C. Gagnon. 2006. Considering the temporal and spatial scales of noise exposures from seismic surveys on baleen whales, IWC/SC/58/E9. Submitted to Scientific Committee, International Whaling Commission. 9 pp, available from the Office of the *Journal of Cetacean Research and Management.*

68. Nieukirk, S.L. *et al.* 2012. Sounds from airguns and fin whales recorded in the mid-Atlantic Ocean, 1999–2009. *J. Acoust. Soc. Am.* **131**: 1102–1112.

69. Laporte, E.D. 2003. Opinion and order granting Plaintiffs' motion for summary judgment. 279 F.Supp.2d 1129 (N.D. Cal.).

70. NMFS. 2012. Taking marine mammals incidental to seismic survey in Cook Inlet, Alaska. *Federal Register* **77**: 27720–27736.

71. Hatch, L.T. & K.M. Fristrup. 2009. No barrier at the boundaries: implementing regional frameworks for noise management in protected natural areas. *Mar. Ecol. Prog. Ser.* **395**: 223–244.

72. Parsons, E.C.M. *et al.* 2009. A critique of the UK's JNCC seismic survey guideline for minimizing acoustic disturbance to marine mammals: best practice? *J. Mar. Poll. Bull.* **58**: 643–651.

73. Dolman, S. *et al.* 2009. *Technical Report on Effective Mitigation for Active Sonar and Beaked Whales.* European Cetacean Society. Istanbul, Turkey.

74. National Research Council. 2005. *Marine Mammal Populations and Ocean Noise: Determining When Noise Causes Biologically Significant Effects.* National Academy Press. Washington, DC.

75. Wood, J., B.L. Southall & D.J. Tollit. 2012. PG&E offshore 3-D Seismic Survey Project EIR – Marine Mammal Technical Draft Report. SMRU Ltd.

76. NOAA. 2012. Cetacean and Sound Mapping. Available at: http://www.st.nmfs.noaa.gov/cetsound. Accessed on Jan. 10, 2013.

77. Agardy, T. *et al.* 2007. A global scientific workshop on spatio-temporal management of noise. Report of workshop held in Puerto Calero, Lanzarote, June 4–6, 2007.

78. Southall, B.L. & A. Scholik-Schlomer. 2008. Final report of the NOAA International Conference: 'Potential Applicatino of Vessel-Quieting Technology on Large Commercial Vessels,' 1–2 May 2007, Silver Spring, Maryland, U.S.A. Silver Spring: NMFS.

79. Harvell, C.D. *et al.* 1999. Review: marine ecology—emerging marine diseases—climate links and anthropogenic factors. *Science* **285**: 1505–1510.

80. Gulland, F.M.D. & A.J. Hall. 2007. Is marine mammal health deteriorating? Trends in the global reporting of marine mammal disease. *EcoHealth* **4**: 135–150.

81. Ward, J.R. & K.D. Lafferty. 2004. The elusive baseline of marine disease: are diseases in ocean ecosystems increasing? *Plos Biol.* **2**: 542–547.

82. Osterhaus, A. & E.J. Vedder. 1988. Identification of virus causing recent seal deaths. *Nature* **335**: 20–20.

83. Grachev, M.A. *et al.* 1989. Distemper virus in Baikal seals. *Nature* **338**: 209–209.

84. Lipscomb, T.P. *et al.* 1994. Morbilliviral disease in an Atlantic bottle-nosed dolphin (Tursiops truncatus) from the Gulf of Mexico. *J. Wildl. Dis.* **30**: 572–576.

85. Schulman, F.Y. *et al.* 1997. Re-evaluation of the 1987–88 Atlantic coast bottlenose dolphin (Tursiops truncatus) mortality event with histologic, immunohistochemical, and molecular evidence for a morbilliviral etiology. *Vet. Pathol.* **34**: 288–295.

86. Rowles, T.K. *et al.* 2011. Evidence of susceptibility to morbillivirus infection in cetaceans from the United States. *Mar. Mammal Sci.* **27**: 1–19.

87. Bossart, G.D. *et al.* 1998. Brevetoxicosis in manatees (Trichechus manatus latirostris) from the 1996 epizootic: gross, histologic, and immunohistochemical features. *Toxicol. Pathol.* **26**: 276–282.

88. Flewelling, L.J. *et al.* 2005. Red tides and marine mammal mortalities. *Nature* **435**: 755–756.

89. Van Dolah, F.M. 2000. Marine algal toxins: origins, health effects, and their increased occurrence. *Environ. Health Persp.* **108**: 133–141.

90. Brodie, E.C. *et al.* 2006. Domoic acid causes reproductive failure in California sea lions (*Zalophus californianus*). *Mar. Mammal Sci.* **22**: 700–707.

91. Harvell, D. *et al.* 2004. The rising tide of ocean diseases: unsolved problems and research priorities. *Front. Ecol. Environ.* **2**: 375–382.

44. Williams, M. 2000. Dark ages and dark areas: global deforestation in the deep past. *J. Historic. Geogr.* **26:** 28–46.

45. Williamson, M., K.J. Gaston & W.M. Lonsdale. 2001. The species—area relationship does not have an asymptote! *J. Biogeogr.* **28:** 827–830.

46. Diamond, J.M. 1972. Biogeographic kinetics—estimation of relaxation-times for avifaunas of Southwest Pacific Islands. *Proc. Natl. Acad. Sci. USA* **69:** 3199–3203.

47. Tilman, D., R.M. May, C.L. Lehman & M.A. Nowak. 1994. Habitat destruction and the extinction debt. *Nature* **371:** 65–66.

48. Kuussaari, M., R. Bommarco, R.K. Heikkinen, *et al.* 2009. Extinction debt: a challenge for biodiversity conservation. *Trends Ecol. Evol.* **24:** 564–571.

49. Ferraz, G., G.J. Russell, P.C. Stouffer, *et al.* 2003. Rates of species loss from Amazonian Forest fragments. *Proc. Natl. Acad. Sci. USA* **100:** 14069–14073.

50. Halley, J.M. & Y. Iwasa. 2011. Neutral theory as a predictor of avifaunal extinctions after habitat loss. *Proc. Natl. Acad. Sci. USA* **108:** 2316–2321.

51. Bierregaard, R.O. & T.E. Lovejoy. 1988. Birds in Amazonian forest fragments: effects of insularization. In *Acta XIX Congressus Internationalis Ornithologici*, Ouellet, H. Ed.: 1564–1579. University of Ottawa Press. Ottawa.

52. Walther, G.-R., S. Beißner & C.A. Burga. 2005. Trends in the upward shift of alpine plants. *J. Vegetation Sc.* **16:** 541–548.

53. Jackson, S.T. & D.F. Sax. 2010. Balancing biodiversity in a changing environment: extinction debt, immigration credit and species turnover. *Trends Ecol. Evol.* **25:** 153–160.

54. Patterson, B.D. & W. Atmar. 2008. Nested subsets and the structure of insular mammalian faunas and archipelagos. *Biol. J. Linnean Soc.* **28:** 65–82.

55. Harte, J., A. Ostling, J. Green & A. Kinzig. 2004. Biodiversity conservation—climate change and extinction risk. *Nature* 430.

56. Gaston, K.J. 2006. Biodiversity and extinction: macroecological patterns and people. *Progr. Phys. Geogr.* **30:** 258–269.

57. Tjorve, E. 2009. Shapes and functions of species–area curves (II): a review of new models and parameterizations. *J. Biogeogr.* **36:** 1435–1445.

58. Pereira, H.M., L. Borda-de-Água & I.S. Martins. 2012. Geometry and scale in species–area relationships. *Nature* **482:** E3–E4.

59. Stouffer, P.C., E.I. Johnson, R.O. Bierregaard & T.E. Lovejoy. 2011. Understory Bird Communities in Amazonian Rainforest Fragments: species turnover through 25 years postisolation in recovering landscapes. *PLoS One.* 6.

60. Myers, N. 1979. In *The Sinking Ark. A New Look at the Problem of Disappearing Species.* Pergamon Press. Oxford.

61. Wilson, E.O. 1992. *The Diversity of Life.* New York: Harvard University Press.

62. He, F. & S.P. Hubbell. 2011. Species–area relationships always overestimate extinction rates from habitat loss. *Nature* **473:** 368–371.

63. Budiansky, S. 1994. Extinction or miscalculation. *Nature* **370:** 105.

64. Lomborg, B. 2001. The skeptical environmentalist: measuring the real state of the world. *Cambridge.* 165–172.

65. J.L. Simon & A. Wildavsky. 1993. Economist vs Biologists on Extinction. *New York Times.* May 13, 1993.

66. Khan, A. 2011. Global species extinction isn't quite so dire, study finds. *Los Angeles Times.* May 21, 2011. http://articles.latimes.com/2011/may/21/science/la-sci-extinction-20110521.

67. Species loss far less severe than feared: study. *DAWN.COM.*

68. Pereira, H.M., P.W. Leadley, V. Proenca, *et al.* 2010. Scenarios for Global Biodiversity in the 21st Century. *Science* **330:** 1496–1501.

69. Pereira, H.M. & G.C. Daily. 2006. Modeling biodiversity dynamics in countryside landscapes. *Ecology* **87:** 1877–1885.

Ann. N.Y. Acad. Sci. ISSN 0077-8923

Ecology and conservation of ginseng (*Panax quinquefolius*) in a changing world

James B. McGraw,[1] Anne E. Lubbers,[2] Martha Van der Voort,[3] Emily H. Mooney,[4] Mary Ann Furedi,[5] Sara Souther,[6] Jessica B. Turner,[1] and Jennifer Chandler[1]

[1]Department of Biology, West Virginia University, Morgantown, West Virginia. [2]Department of Biology, Centre College, Danville, Kentucky. [3]Department of Biology, New Mexico Highlands University, Las Vegas, New Mexico. [4]Department of Biology, Massachusetts College of Liberal Arts, North Adams, Massachusetts. [5]Western Pennsylvania Conservancy, Middletown, Pennsylvania. [6]Department of Botany, University of Wisconsin, Madison, Wisconsin

Address for correspondence: James B. McGraw, Department of Biology, P.O. Box 6057, West Virginia University, Morgantown, WV, 26506-6057. jmcgraw@wvu.edu

American ginseng (*Panax quinquefolius* L.) is an uncommon to rare understory plant of the eastern deciduous forest. Harvesting to supply the Asian traditional medicine market made ginseng North America's most harvested wild plant for two centuries, eventually prompting a listing on CITES Appendix II. The prominence of this representative understory plant has led to its use as a phytometer to better understand how environmental changes are affecting many lesser-known species that constitute the diverse temperate flora of eastern North America. We review recent scientific findings concerning this remarkable phytometer species, identifying factors through its history of direct and indirect interactions with humans that have led to the current condition of the species. Harvest, deer browse, and climate change effects have been studied in detail, and all represent unique interacting threats to ginseng's long-term persistence. Finally, we synthesize our current understanding by portraying ginseng's existence in thousands of small populations, precariously poised to either escape or be drawn further toward extinction by the actions of our own species.

Keywords: ginseng; *Panax quinquefolius*; deer browsing; climate change; harvest; extinction vortex

Introduction

The understory herb layer constitutes, on average, more than 80% of the total plant species in forest communities.[1] These accessible ground-layer plants serve as resources for animal mutualists, herbivores, predators, fungal mutualists, diseases, and a diverse microbial community. As such, the long-term fate of herbaceous plants largely determines the overall biodiversity trend in forests. Extinction of forest herb populations, locally or globally, must be balanced by immigration of new species, or overall forest community diversity will decline.

In order to gauge the threats to forest herbaceous plant populations, a bottom–up approach would involve extensive instrumentation of forest understory environments to measure the changes in those environments over time and space. Although it would be inaccurate to describe Project NEON (National Environmental Observation Network)[2] as solely an instrumentation exercise, such efforts represent a large percentage of the project's focus. The ecological interpretation of data gathered in this way presupposes detailed knowledge of how organisms will respond to changes in the observed environment.

Long ago, Clements and Goldsmith[3] suggested a top–down alternative—the *phytometer* approach—asserting that the best way to study the environment is to measure the response of the plant itself. At the time, well-characterized crop plants were used to measure environmental effects. However, in the past few decades, a strong case has been made for using native plants to measure environmental effects, particularly in natural settings.[4,5] In the past three decades a small community of ecologists has focused their attention on one particularly well-known understory herb—the medicinal plant, *Panax quinquefolius* L.—hereafter referred to as American

doi: 10.1111/nyas.12032
Ann. N.Y. Acad. Sci. 1286 (2013) 62–91 © 2013 New York Academy of Sciences.

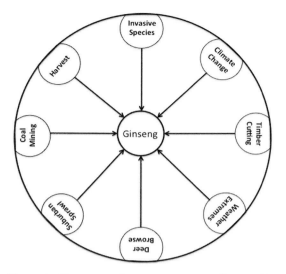

Figure 1. Ginseng as a phytometer to measure direct and indirect effects of human activities and policies in the forest environment (simplified and modified from Ref. 6).

ginseng, or just ginseng (Supporting Fig. S1). Members of our laboratory have referred to these studies, collectively, as Project GEON (Ginseng Environmental Observation Network) (Fig. 1).

On the face of it, selecting a medicinal plant harvested from the wild as the representative understory plant species for ecological study would seem to be a serious mistake. However, the harvesting of ginseng motivated the listing on CITES Appendix II in 1975, which, in turn, prompted early studies of its population growth and ecology.[7–11] These studies demonstrated that ginseng was a long-lived, widespread perennial plant, and that it exhibited the slow life history typified by forest understory plants generally, rendering it suitable for demographic studies and for the use of both individuals and populations as phytometers. The most atypical characteristic of ginseng was the appetite and accompanying reverence of Asian cultures for the curative powers of the twisted, gnarly storage root.[12] The economic demand driven by this appetite results in the wild ginseng harvest. The harvesting of ginseng adds direct human interaction to the set of factors influencing the plant, which, in turn, means that scientific research on the species has a salience for the public that might be lacking for a less well-known species.

One purpose of this review is to take stock of the expanding scientific literature on American gin-

seng ecology and conservation biology in order to identify factors likely to influence the long-term fate of the species, and by proxy, other herbaceous understory plants. We emphasize the scientific literature because of the large circulating body of unsubstantiated information concerning this plant, which ranges from folklore to presumption to repeated but unquantified observations by nonscientists. By focusing on scientific findings, we identify aspects of this tradition that are supported by evidence. While we occasionally refer to the literature on cultivated, woods-grown or wild-simulated ginseng growing approaches, our emphasis here is on the ecology and conservation of natural populations. This leads to the second purpose of this review; to identify critical gaps in our understanding that could ultimately lead to better management of forest understory species and extinction prevention.

Brief history of the ginseng–human relationship

In North America, ginseng was considered a botanical resource of minor importance for the Native American apothecary,[13] although this varied widely among tribes.[14] In the early 1700s, with assistance from Native Americans, members of the Jesuit order confirmed the taxonomic relationship of *P. quinquefolius* to *P. ginseng*, a relative on the Asian continent that had been revered for its medicinal properties for millennia. *P. ginseng* had become extremely rare in Asia, most likely due to overharvest and deforestation.[12] American ginseng, though considered by the Chinese to possess different "powers" than Asian ginseng, was nevertheless valued for its medicinal properties. Thus when *P. quinquefolius* was discovered by Jesuits in North America, the market was sufficiently profitable to stimulate intense wild harvest, eventually reaching an industrial scale.[13] Reports of dried root exports in huge quantities suggest much larger natural population densities than are observed today. For example, in one typical year (1841), more than 290,000 kg of dry ginseng roots were shipped from North America to the Asian continent. Although average root size was larger in the 1800s than it is today,[15] even a conservative estimate would suggest that this represents at least 64 million roots. The total annual harvest during the 1800s was therefore approximately an order of magnitude greater than the wild harvest in the most recent decade (2001–10; USFWS 2011, CITES nondetriment finding).

The relative scarcity of wild ginseng by the end of the 1800s led to intense efforts to cultivate the plant to satisfy continuing market demand.[13] These efforts frequently met with failure, though the reasons for this were not always clear. Grown at high densities, both root and leaf fungal diseases wiped out many crops, often after multiple seasons of cultivation and expense. Theft was also a problem given the high value of the root. Inadequate soil or climatic conditions affected other experiments in cultivation. Volatile market prices confounded entrepreneurial farmers who invested hundreds of dollars, hoping to cash in on the bonanza. Finally, roots produced under cultivation were typically not the twisted gnarly phenotype prized by Asian buyers, and fetched a lower price. Despite these problems, a small cadre of growers, particularly in Wisconsin, Ontario, and much later in the dry valleys of British Columbia, succeeded in commercializing ginseng farming. The price differential with the crop's wild ancestor drove two separate harvest systems that continue to the present.

American ginseng was listed on Appendix II of CITES (Convention on International Trade in Endangered Species of Wild Fauna and Flora) in 1975. All species listed on Appendix II are considered to be susceptible to extinction in the absence of trade controls. Trade of any species listed on Appendix II requires a permit and all species are subject to an annual "no detriment" finding in order to maintain permitted trading status. Ginseng was listed due to concern over existing high levels of harvest.[16] The CITES ginseng program is managed at the federal level by the U.S. Fish and Wildlife Service.[16] USFWS required states to implement a ginseng regulatory structure, including harvest seasons, recordkeeping regarding sales, and a set of rules for harvesters. In the early years of CITES regulation, harvest seasons varied widely among states; however, range-wide studies showed no biological basis for this variation.[17] Most states are converging toward a uniform start date of September 1, which increases the possibility of planting ripe seeds that will allow recovery of the population. Record-keeping requirements are modest, primarily involving the recording of sales events to facilitate monitoring of total harvest as well as per root biomass (see Ref. 16). Harvest and permitting requirements vary on state and U.S. Forest Service Land. Harvest on private land does not generally require a permit; however, diggers must secure permission from the landowner. Federal law requires that all wild-harvested ginseng plants are at least five years of age.[18] Most states also strongly suggest or require that all seeds of harvested plants be sown *in situ* at the harvest site.

Geographic distribution

American ginseng historically was found in rich, cool hardwood forests extending from southern Québec and Ontario, south to northern Georgia, and west as far as Minnesota, eastern Oklahoma, and northern Louisiana.[19] Throughout this wide range, however, natural populations vary from frequent to uncommon to rare across the landscape, but they are almost always small, generally being made up of fewer than 200 individuals.[20,21]

Contrary to conventional wisdom, the species does not appear to have very specialized habitat requirements. While populations may be more frequently encountered in mixed mesophytic forests with canopy species such as maple, tulip poplar, basswood, and black cherry, they can also be found in drier oak-hickory dominated forests, or moister sites with a walnut or sycamore canopy, though they are rarely encountered in floodplain forests or the most xeric, south-facing, and ridgetop sites. Populations also occupy sites ranging from flat to steep hillsides, and at least in the middle of its range, on all aspects. While populations in the south tend to be found in the mountains, in the north they are at lower elevations, and in the middle of the range, elevations vary from 200 to > 1000 meters.[10,20–23] Soils supporting ginseng populations are also variable, ranging from moderately acid to near neutral and having textures ranging from loamy sand to silty clay.[10,20,22] In Arkansas and Indiana, A-horizon levels of phosphorus and calcium were similar to those of sites hospitable to other forest species, although potassium levels were low.[20,22] Variation in soil pH, slope, aspect, and elevation among 30 representative natural populations across seven states is shown in Figure 2. Given these facts, ginseng can be considered to have a most uncommon sort of rarity;[24] small populations, large geographic range, and a broad niche.

Within populations, the dispersion pattern is highly clumped,[25,26] with cluster size ranging greatly from isolated individuals with no plants in a 3 m

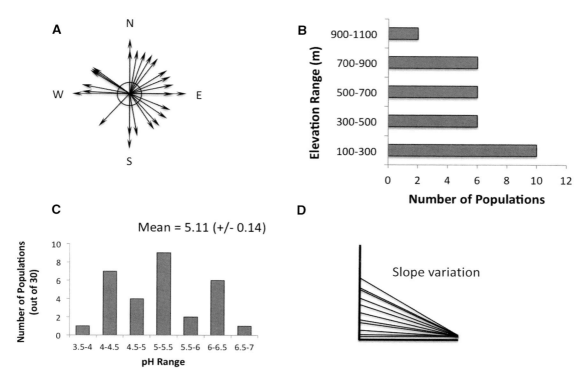

Figure 2. Range of (A) aspects, (B) elevations, (C) soil pH, and (D) slopes occupied by 30 representative natural populations of American ginseng.

radius to >100 plants within a 1 m² area.[27] For one typical large population (total $N = 369$) censused in 2011, there were 50 clusters with a mean cluster size of 7.38 plants/cluster (range; 1–28 plants). Genetic analyses have suggested such clusters to be, at least in part, family groupings resulting from limited seed dispersal.[25]

Life history

Patterns of growth

The basic morphology, phenology, and overall life history of ginseng have been well documented.[7,8,10,20] The perennating organ consists of a fleshy taproot supporting a short underground rhizome. Additional root structures develop either from the side of an existing root or from the rhizome, leading to the typical multibranched storage organ prized by harvesters. An aboveground stalk (sometimes referred to as a sympodium) is attached to the apical end of the rhizome, and consists of fused leaf petioles and peduncle. In plants with more than one leaf, the petiole of each leaf branches laterally from a point at the stalk apex, while the peduncle

grows vertically from the center. Stalks can support up to four (rarely five or six[10]) palmately compound leaf blades and potentially one umbellate inflorescence. Each leaf is referred to as a prong, a colloquial ginseng-specific term. Aboveground growth is determinate; leaves and flowers lost cannot be replaced within a growing season.

Because each senesced annual stalk leaves one scar on the rhizome, plants can be aged by counting the scars on the rhizome.[28] Age is positively, and nonlinearly, related to leaf area and stalk height; the relationship, however, varies among individuals, among microsites within populations, as well as among populations.[29] Figure 3 illustrates the variability in the age–size relationship for three large natural populations. Young plants are small and exhibit low variance in size, but as they age, plant size variation increases dramatically, probably reflecting many kinds of factors, both environmental and genetic, that cause variable growth rates. As is true for most plants, size is a better predictor of flower and fruit numbers than age.[8,20,29] Although estimates of maximum lifespan vary from 25 to 30 years[20] to

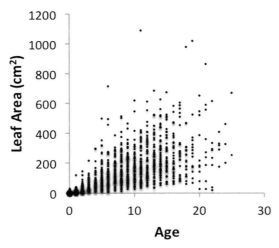

Figure 3. Relationship of age and size (measured as leaf area) in three natural populations of American ginseng.

more than 50 years,[8,10] in most natural populations few individuals live more than 25 years.[29] Harvesting in a given population removes the larger and often older individuals, obscuring accurate determination of potential lifespan. Unharvested older individuals likely die from a variety of causes, most of which are poorly documented. Indeed, the cause of mortality as plants age is one of the greatest unknowns regarding ginseng life history.

The morphology of ginseng allows delineation of a series of discrete life stages:[7,8,10,20] seeds, first-year seedlings, older one-leaf seedlings, two-leaved juveniles, and three- to five-leaved adults. Seedlings typically possess one leaf with three leaflets, although four or five leaflets may occur in the second or later years. In natural populations (vs. cultivation) plants may remain in the one-leaf seedling stage for two to five years or occasionally longer. For juveniles and adults, transition to a larger leaf number-class does not occur on a yearly basis, and stasis or retrogression to a lower leaf number is not unusual. Within a given leaf number class, undisturbed plants often grow from year to year as measured by number (3–5) and size of leaflets on each leaf. Once plants reach the two-leaved stage, flowering is possible.

The annual phenology of *P. quinquefolius* conforms to that of a summer green geophyte[30] whose leaves mature synchronously along with the overstory canopy, and whose senescence is variable. In Wisconsin and Illinois, juvenile and adult plants emerge in April–May, with inflorescences starting to develop slightly later than the leaves.[7,20] Leaves fully expand about one month after plant emergence. Flowers mature centripetally beginning in mid-June to mid-July.[31] Fruit ripening within an inflorescence is also asynchronous;[17] mature, reddened fruits are observable in certain sites and years by late July,[7,20] although across the range fruit ripening more typically occurs in late August to early September.[17] Fruit dispersal continues to mid-October and is followed directly by plant senescence. Seedling and juvenile plants tend to senesce earlier than adults.[7] By senescence the rhizome has developed buds containing leaf and floral primordia for the following year.[20,32] For a more detailed anatomical description of the phenological stages of germination and plant development see Ref. 33. Specific dates of these phenological stages vary geographically with elevation, aspect, latitude and year.[17,20] At least some of this variation reflects the temperature and precipitation regime experienced by the population.

Seed dispersal in time and space

Ginseng seeds require at least 21 months (two winters) after dispersal in the fall for after-ripening due to their deep simple morphophysiological dormancy,[34] a syndrome common among forest herbaceous perennials.[35] Experiments using commercially obtained seed show that embryo growth is minimal for the first 9–10 months after dispersal (September–mid-May), but over the next 4–5 months cotyledons grow rapidly and the endocarp softens and splits.[36] Stratification temperatures of 20° C are best at promoting such development.[37] Once the morphological component of dormancy is broken, a physiological dormancy mechanism prevents the seeds from germinating until after their second winter.[35] Seeds occasionally have been observed to germinate only nine months after dispersal, that is, during their first spring, but this is rare.[37,38] Seeds from red, ripe fruits are much more likely to germinate than seeds from green fruits.[17]

While some seeds are capable of germinating 21 months after dispersal, others may persist in the soil for three, four, or even five winters after dispersal, depending on site characteristics.[39] This has led to an understanding of the age-structured seed bank that builds in a time-delay for germination ranging from 21 to 45 months. Although modeling experiments have demonstrated that population growth rates are relatively insensitive to this time-delay, the seed bank could nevertheless play an important role in recovery from harvest and persisting through poor environmental conditions.[40]

Most fruits drop close to the parent plant[8] similar to many forest herbs,[41] with the potential to be later cached by small mammals such as mice or chipmunks.[42] The red, fleshy drupes have the hallmarks of fruits that are dispersed by birds, and if seeds remain viable within the bird digestive system, this could provide an opportunity for long-distance dispersal. Consumption rates by animals and the accompanying dispersal distances are not known for natural populations, but could play an important role in metapopulation dynamics as well as migration in response to environmental change since dispersal by gravity is typically limited.

Seedling ecology
Consistent with findings of broad distribution, seedling emergence, seedling survival, and the consequent net recruitment rate were not significantly affected by aspect.[23] Leaf litter depth, however, had a large effect: shallow (ambient) litter increased all three measures of seedling success relative to bare soil.[23] Following establishment, light regime may play an important role in growth. Fournier *et al.*[43] related growth of seedlings to diurnal and seasonal variation in the light environment of a deciduous forest canopy in Québec. Mean daily sunfleck durations of up to two hours positively affected biomass, explaining up to 56% of the variation in root and shoot dry mass. For plants experiencing sunfleck durations of more than two hours per day, diffuse and direct photon flux density had a positive impact, and explained 69% of the variation in shoot mass and 52% in root mass, leading to the conclusion that light is an important limiting factor for much of the summer growth period.[43] Summer green geophytes have been shown to respond more rapidly to sunflecks than spring green herbs.[44] They also acclimate to the lower irradiance environment after canopy closure through reductions in light compensation point,[45] although few measures of photosynthetic properties of ginseng have been made.[46]

Reproductive biology
Ginseng flowers have five sepals, petals, and stamens, and a single inferior ovary with one to three ovules.[32,47] Ovule number can be determined by style number, although styles do not separate until anther dehiscence.[31,32] Most flowers begin with

two ovules, but some of them abort an ovule very early and the two styles remain fused.[47,48] Individual ovules may not get fertilized, or they may abort after fertilization. The number of seeds produced by a single fruit can be readily determined by counting lobes. Structurally, the fruits are berry-like drupes,[8] and are often referred to as berries.

Commonly reported visitors to the flowers are small bees (Halictidae: *Lasioglossum*) and flies (*Syrphidae*).[7,49] Species in both of these groups have been found to carry pollen,[8,31,50] with larger pollen loads collected from the bees.[50] Visitation rates tend to be low (e.g. three or fewer visitors in 30 minutes of observation[31]). Other types of insects observed to carry pollen in low amounts include bugs (*Lygus lineolaris*), ants (*Lasius* sp.) and flies (*Delia* sp.).[50]

Although ginseng does not reproduce asexually,[20] production of two stalks from one rhizome occurs occasionally.[8] Experimentally planted pieces of root and rhizome can regenerate plants by activating adventitious vegetative buds.[51] Several studies have documented the mixed-mating system of the species.[7,28,31,38] The absence of viable seed production from bagged, emasculated plants suggests that the species is not capable of apomixis.[31] However, with anthers left in place, seed production occurs, demonstrating that self-pollination is a component of the breeding system.[7,31] On the other hand, at least some of the genetic diversity of populations is due to outcrossing.[52,53]

Selfing had been thought to be solely *via* geitonogamy because the flowers were reported to be protandrous;[7] however, the degree of protandry varies among individuals and populations, with at least some flowers having simultaneous stigma maturation and anther dehiscence.[28] The presence of pollen tubes before the anthers dehisced and the styles separated in 27% of hand-pollinated flowers led Schlessman[31] to conclude that autogamy is possible, though pollen tubes were most prevalent in flowers pollinated after the anthers dropped off. Schluter and Punja[47] concur that autogamy is likely, but for the reason that anther dehiscence is staggered within a flower, and anthers are still present when stigmas separate. They did find, however, that pollen germination and pollen tube growth occurred only in flowers whose styles had started to separate; they suggested that because stigma separation was not always obvious at the time of flower collection, Schlessman[31] may not have been able to tell that it had already begun when he observed the pollen tubes. Geitonogamy remains probable as well, due to the proximity of flowers in the inflorescence. The level of geitonogamy will be restricted by the fact that the flowers of an inflorescence open centripetally over a one- to four-week period[7,31,47] and no more than eight flowers[28] or 10% of flowers[47] in an inflorescence were open at any one time.

At a fundamental level, reproductive success can be attributed in a multiplicative manner to its components: probability of producing an inflorescence, the number of flowers per inflorescence, the proportion of flowers producing fruit, and the number of seeds per fruit. The first two components tend to increase with plant size.[7,8,31,47] In six Kentucky populations, the probability of producing an inflorescence in 2-leaf plants was 40–90%, whereas three- and four-leaf adults did so 90–100% of the time. Reproductive two-leaf plants had a mean of 5–10 flowers per inflorescence, while reproductive adults had mean flower numbers of 8–43 flowers per plant, with the mean varying considerably among populations. Also, in the six Kentucky populations, the proportion of flowers that matured into fruit did not vary among size classes and likewise seeds per fruit did not vary among size classes.

In very small populations or isolated clusters, indirect evidence for pollinator limitation was found in experimental populations, resulting in an Allee effect.[54] However, even when pollination is facilitated, seed set may be well below its potential.[38,45,47] Schluter and Punja[47] observed pollen tube growth in the persistent styles of fruits that had been aborted. Seed set rose from 38% in flowers from intact inflorescences to 48% when most of the flower buds had been removed.[31] At least some of the reduction in seed set below potential is likely due to internal and external resource limitation. Internal resource limitation is reflected in the positive relationship between age or size and the proportion of flowering plants that produce fruit, as well as the total numbers of fruits and seeds produced.[7,8,20,29]

Demography

Several life history properties of ginseng lend themselves to demographic studies. Distinct stages with contrasting survival, growth, and reproductive properties allow the population to be logically divided into classes. The static aboveground size of plants within a year means that precise census timing is not required to assign individuals to classes (vis-à-vis plants that grow continuously through the season). The nonclonal nature of individuals makes counting genets straightforward. Precise counts of predispersal seed numbers make fertility calculations easy. Large, visible seeds that have limited dormancy make incorporation of the seed bank into demographic models relatively easy with supplementary seed cage studies. Finally, relatively small population sizes mean that censuses can be readily performed for entire populations.

The first formal attempt to model population dynamics of ginseng was carried out by Charron and Gagnon[10] for four populations near the northern edge of the range in Québec. This study ignored age structure in the seed bank, and the five remaining stage classes were determined by leaf number, with one-leaf new seedlings being distinguished from older one-leaf seedlings. Charron and Gagnon[10] compared plant size and age as predictors of plant survival, growth and reproduction, concluding that size was the more important arbiter of performance. Therefore, a 6 × 6 stage–based population projection matrix was assembled for each population; two populations for three transition years and two for one transition year. Despite limited sample sizes ($60 \leq N \leq 132$), the transition probabilities in these matrices established that ginseng had the characteristic slow life history common to many understory herbaceous plants. Survival, growth, and reproduction varied sharply among classes with the seed and seedling stages being the most vulnerable. A long prereproductive period was evident as many one-leaf and two-leaf plants remained in the same class from one year to the next. Adult stage fertilities were low (40 seeds per four-leaf adult being the highest observed), but later comparisons with other studies suggest these are actually relatively high fertilities compared to central or southern populations.

The dominant eigenvalue of the transition matrix yields the finite rate of increase, λ. Charron and Gagnon concluded that λ was near 1, indicating stable population sizes when at the stable stage distribution, again similar to what has been observed for other understory herbaceous plants. Several other studies of ginseng population dynamics have employed matrix population models to estimate λ as a synthetic measure of population performance in the context of studies of harvest,[26,56] deer browse,[57,58] both harvest and deer browse,[59]

optimally plants ginseng seeds 2 cm deep at the time of harvest, can sustain population growth rates equal to that of unharvested populations.

Farrington *et al.*[59] conducted demographic simulations that took into consideration the interacting effects of ginseng harvest and white-tailed deer (*Odocoileus virginianus*) browse. Harvest and browse had negative effects on the population dynamics of ginseng; however, the effects were nonadditive. The presence of browse reduced λ, but only marginally; λ was still > 1.0. LTREs demonstrated that deer browse had a positive effect on ginseng survival because browse actually protected large plants by making them invisible to harvesters. Farrington *et al.*[59] also found that "responsible" seed planting (i.e., seeds planted at a depth of 2 cm) at the time of harvest resulted in higher growth rates than plants harvested "irresponsibly" (no seed planting).

Observed effects of harvest on populations. Recovery of two well-characterized small populations of ginseng from severe harvest has been documented. Lewis[9] monitored a population in Missouri that was decimated by harvest the year after plants were thoroughly censused. In fall of 1996, all visible ginseng plants were removed in a complete experimental harvest in West Virginia.[51] Both studies showed rapid recovery of the total aboveground population size (Fig. 5A for West Virginia); however, stage structure was much slower to recover. Lewis[9] deduced that the seed bank must have been responsible for the early regrowth of the Missouri population, and in West Virginia, the same pattern was seen.[51] In Missouri, after five years, only 25% of plants in the population were reproductive versus 66% before harvest. In West Virginia, the adult portion of the population also recovered slowly but steadily for the first decade after harvest (Fig. 5B).

This resulted in seed production and a pulse of recruitment in 2008. An unknown event resulted in additional adult mortality, such that the adult population remains far below preharvest levels. The built-in resistance to harvest through the presence of a seed bank left by older, more fecund individuals may not be present in populations with reduced fertilities.

Harvest effects: genetic. Wild harvest can affect the genetic diversity and evolution of targeted species, as shown by numerous examples from fisheries, forestry, and game management.[65–68] The results of this interaction between humans and wild species are frequently reduced genetic diversity, unintentional selection against desirable traits, and ultimately reduced population viability. The introduction of individuals to reverse population declines also has genetic consequences, namely outbreeding depression—reduced fitness resulting from introduction of maladapted alleles or the breakup of coadapted gene complexes.[69] As with other wild-harvested species, these genetic issues are cause for concern for ginseng conservation.

Harvest of plants from the wild can clearly affect patterns of genetic diversity.[65,70] Because harvest acts as an independent bottleneck event, each population would have a unique set of their alleles lost or fixed, resulting in a high level of interpopulation differentiation combined with low levels of intrapopulation diversity for a targeted species. This impact of harvest is consistently observed in patterns of genetic variation in ginseng, although these have been assessed with a variety of sampling regimes, marker systems, and statistics.[25,52,53,71–75] Genetic variation within and among contemporary populations of ginseng is likely influenced by harvest history, but this may also be exacerbated by

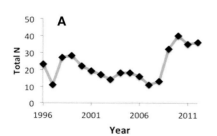
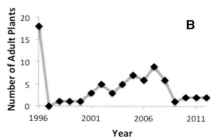

Figure 5. Recovery of (A) the total population, and (B) the adult (three-leaved) component of the population after experimental harvest in 1996 for a small population in northern West Virginia.

life-history characteristics that reduce gene flow. Within populations, plants within 2 m of one another tend to be genetically related;[53] this level of genetic structure is the result of limited pollen movement and/or seed dispersal.[76] While less is known about seed dispersal, the largely autogamous breeding system of ginseng would limit pollen movement. Interpopulation genetic differentiation estimates (G_{ST}) for ginseng range between those typical of plants with mixed mating systems and those typical of inbreeding dicots.[77] Limited pollen or seed movement would reduce gene flow among populations and thereby increase divergence.[77]

The consequences of harvest for genetic diversity in ginseng populations have been determined by both modeling and field studies. A simulation study conducted by Cruse-Sanders *et al.*[78] found that harvesting as few as 10% of mature (>2 leaves) plants in populations significantly reduced genetic diversity. The results from the simulation study were supported by comparisons of protected populations to those legally open to harvest.[53] Levels of expected heterozygosity (H_e) were significantly lower in unprotected populations ($H_e = 0.070$) than in protected populations ($H_e = 0.076$). Unprotected populations showed evidence of recent bottleneck events, while no such events were detectable in any protected populations. In contrast, Obae and West[79] did not find reduced genetic variability in populations in high harvest versus low harvest regions, but this was likely due to a small sample size ($n = 4$) in the high harvest region.

Indirect genetic effects result from small population sizes and low levels of genetic diversity following harvest. Chief among these effects is increased inbreeding. Studies from other species suggest that reduced density of conspecifics would increase the proportion of seeds produced by self-pollination, and when outcrossing takes place, reduced genetic diversity would mean a greater likelihood that nonself-pollen would be genetically similar.[80] In fact, results from allozyme analyses of ginseng suggest that offspring in wild populations are largely the products of inbreeding.[25] Such mating between close relatives may result in reduced fitness in offspring, that is, inbreeding depression. Alternatively, the deleterious recessive alleles primarily responsible for inbreeding depression may be purged from populations with histories of inbreeding. Successive generations of inbreeding will expose alleles to elimination by natural selection, and therefore, inbreeding may not lead to inbreeding depression in these species. However, a meta-analysis by Byers and Waller[81] suggests that purging is inconsistent among plant species, and substantial decreases in offspring fitness are possible even for predominantly self-pollinating species.[82] This is especially likely where inbreeding has increased only recently in the history of a population, as is likely in ginseng since evidence suggests populations were much larger only 300 years ago. One study directly examined the consequences of inbreeding for ginseng in three natural populations.[38] Relative to those produced by cross-pollination, offspring of self-pollination had reduced stem height and leaf area, the latter of which is a trait positively correlated with longer term survival.[38] Thus, a shift in breeding system toward more mating among relatives and greater selfing due to smaller population sizes will reduce fitness.

Anecdotal accounts have reported the planting of seeds from cultivated populations into wild populations by managers and harvesters.[53,74,75] Assuming these seeds germinate and grow into adult plants, subsequent outcrossing with wild plants could produce offspring with traits more similar to cultivated plants.[38] The degree to which populations in the wild derive from cultivated seed is an ongoing debate in the management of ginseng.[18] At least six separate studies have applied genetic markers to this question.[52,71,72,74,75] Three of these studies purported to show evidence of genetic introgression of cultivated genotypes into wild populations.[52,74,75] These studies used different molecular markers and sampling protocols, some of which were more robust than others. Boehm *et al.*[52] found that one of the 14 wild populations they sampled was genetically similar to a cultivated source. However, this population was actually a small number of transplants ($N = 15$) to a woodland garden in Pennsylvania. Also using RAPD variation, Schlag and McIntosh[74] found evidence that one of seven wild populations in Maryland contained plants genetically similar to a cultivated population derived from out-of-state seed sources. Harvesters provided the genetic source material for the assessment by Schlag and McIntosh; thus the conclusion that the original populations were wild is circumstantial. Relative to a truly random sample of wild populations, the approaches of these two studies would increase the likelihood of sampling a population supplemented with cultivated seeds by harvesters.

Young and coauthors[75] analyzed a much larger sample of plants from wild and cultivated sources ($N = 489$) using microsatellite markers. They found that some of the wild-collected samples were genetically similar to a Wisconsin/cultivated group. They interpreted this result as suggestive of widespread planting of cultivated seed in natural settings. However, it is not clear how they sampled or identified wild populations in any state, or how many wild-collected samples fell into this group. Grubbs and Case[72] published the most comprehensive and transparent study that addresses genetic differences between wild ($N = 31$) and cultivated ($N = 12$) populations throughout the range of ginseng. Unlike other assessments, the majority of wild populations were located and sampled by state and independent botanists, rather than harvesters. They found that one unique allele (Idh2) was found exclusively in wild populations and there was no evidence of introgression of cultivated genotypes into wild populations. However, showing that introgression of genes from cultivated plants into wild populations has occurred is complicated by several factors. Many cultivated populations have been recently derived from wild populations and they are often the products of independent accessions of wild-derived roots and seeds.[52,83,84] The studies to date have either lacked the rigorous sampling[71,74,75] or the fine-scale genotyping methods[85] to conclusively detect introgression. In summary, it is likely that harvesters are introducing cultivated seed into some wild populations, but the majority of ginseng populations in the wild are not products of this practice. The issue is important as CITES is specifically concerned with protection of wild populations.

Harvest can alter the evolution of ginseng by directly changing the relationship between phenotypic traits and fitness, that is, the pattern of phenotypic selection. In a simulation study, human harvesters selectively removed larger adult plants, which were likely more apparent in the dense forest understory.[26] The outcome of this process is that larger plants lose their fitness advantage, and this effect is exacerbated when harvesters remove seeds. In *Saussurea laniceps* (Himalayan snow lotus), such size-selective harvest has led to the dwarfing of plants in the wild.[86] Two studies suggest that similar dwarfing may be happening in ginseng.[15,29]

The overall size of herbarium specimens has declined over the last century, and this effect was most pronounced among specimens collected from geographical areas with high levels of harvest.[15] In the wild, plants from populations with a high harvest index—defined as the proportion of seedlings and juvenile plants—had smaller leaf areas and stem heights than plants of the same age in populations with a low harvest index.[29] Environmental variation could also be a factor in these results, although these changes are consistent with human-induced evolution in many other species.[67]

Deer browse effects

In addition to human harvest, browsing by white-tailed deer (*O. virginianus* Zimm.) is a pressure likely contributing to the decline of American ginseng populations. As a result of strict hunting regulations, land-use changes, and loss of top predators, white-tailed deer are now the most abundant wild ungulates in North America. According to archeological evidence of deer consumption rates by Native Americans and early European settlers,[87] deer densities are currently two to four times higher than presettlement densities in much of the United States.[88,89] As keystone herbivores in the eastern deciduous forest, large deer herds can alter biotic communities within forest ecosystems.[90–94] The building body of evidence suggests that deer are negatively affecting many aspects of the forest community by depressing the growth of valuable tree[93,95,96] and herbaceous species[97–102] and altering species richness and abundance.[89,95,96,98,103,104] Several studies have documented the effects of deer browsing on American ginseng.[57–59,105] White-tailed deer may be exacerbating the rarity of ginseng, as herbivory, within some populations, occurs at high rates and deer are seed predators of ginseng.[105]

Though human harvesters and deer both remove plant tissue, harvesters remove the root, which results in death of the plant, and they may (or may not) plant seeds to encourage reproduction. Deer effects are likewise variable, but generally affect only aboveground plant parts: deer may remove a portion or all of the leaves, reproductive structures and stalk (Fig. 6), but typically leave the root intact. The loss of aboveground biomass may temporarily remove the plant from a population in the season in which browsing occurs but, as a perennial herb, the potential for regrowth in the following season

Figure 6. Game camera image sequence showing browsing of ginseng by a young white-tailed deer. In this instance, the leaves were consumed but the infructescence was left intact.

remains since the ginseng root is not damaged. Deer may also hinder new recruitment in a population by consuming fruits, since seeds are destroyed during the digestive process.[105]

Deer browsing affects the size distribution and fertility of plants in a population. In studying browse rates and patterns of browse in natural and experimental ginseng populations, Furedi[57] found that plant characteristics and microsite conditions related to apparency influence browse susceptibility. Larger plants (i.e., plants with a greater leaf number, leaf area, and stalk height) were more likely to be browsed than their smaller, shorter counterparts. Similar results were reported by Farrington *et al.*[59] Reproductive plants were more susceptible to browse than nonreproductive ones. Plants in open, unprotected areas were more likely to be browsed than those hidden by fallen logs, rocks, and shrubs and located farther from deer trails. Other studies have reported that food choices made by deer result in morphological changes in plant populations.[91,93,95,96]

By following the fate of individual plants in natural ginseng populations over multiple consecutive growing seasons, Furedi[57] found that negative effects of browse were carried over into the following year. Generally, relative growth rate of leaf area and stalk height were reduced, an effect that was further compounded with consecutive years of browse. Reproduction in the year following browse was also reduced either by the production of fewer buds per inflorescence (an effect further exacerbated by two years of browse) or the absence of a reproductive

structure altogether. Although mortality from deer browsing is difficult to differentiate from dormancy, browsing was associated with nonemergence in the year(s) following browse. Given that true whole plant dormancy is thought to be rare or nonexistent, most of the absences were probably mortality. These patterns are consistent with effects of ungulate herbivory reported in other plant species.[59,97,99,106,107]

The collective negative effects of deer browsing were integrated by determining effects on λ and partitioned with an LTRE. Furedi[57] and McGraw and Furedi[58] showed that the overall population growth rate for seven populations in northern West Virginia was 0.973 (i.e., declining by 2.7%) in the presence of deer browsing but that the same populations would have increased by 2.1% annually ($\lambda = 1.021$) with the removal of the direct effect of browse. The LTRE showed that lowered values for growth transitions of juveniles and small adults were together responsible for about one half of the reduction in λ caused by deer browsing.[57] Reduction in the proportion of large adults staying large adults contributed to one fourth of the reduction in λ. Finally, reduced fertility of large adults explained the remaining one fourth of the λ difference between browsed and unbrowsed populations.

Using population viability analyses (PVA), McGraw and Furedi[58] expanded on the demographic work to examine the effects of current browse levels on stochastic population projections over the next 100 years. Given the current browse rates, the minimum viable population size was calculated at approximately 800 individuals, much higher than

persisted in the year following the frost event.[127] Stochastic events, though infrequent, can have lasting and dramatic consequences for demography.[127] Changes in the frequency of such events may be as important a determinant of ginseng persistence as mean changes in temperature, precipitation, and CO_2 concentration.

Biotic factors, such as incidence of disease, herbivory rate, and competition regime, will be altered as the climate changes.[128] Some of these factors, disease in particular, likely covary with temperature, and thus are implicitly included in the projection of ginseng response to warming mentioned above. Despite this, biotic interactions influence abundance and distribution of species and will no doubt shape response to climate change.[128,129] As global temperatures exceed historical climatic variation, biotic interactions—complex and thus difficult to predict—will affect the fate of species, like ginseng, in unanticipated ways. While such ecological surprises add uncertainty to projections of ginseng response, observational and experimental findings strongly suggest that increasing global temperatures will have a pervasive, negative impact on ginseng populations.[39,46,118,127]

Landscape-level change

The deciduous forests of the eastern United States have been subjected to substantial changes in forest composition, forest cover, and land-use since European colonization began.[130,131] As a species almost exclusively found in the understory of the eastern deciduous forest, *P. quinquefolius* L. has undoubtedly been influenced by these alterations. In the following sections, we briefly describe the nature of these changes and their potential impacts on understory plants, such as ginseng.

Presettlement landscape
Presettlement forests in the eastern United States spanned over 300 million hectares.[130] Community characteristics of mixed-mesophytic old growth forests included decomposing large logs, canopy gaps, multiple vegetation layers, a diverse herbaceous understory, and soil rich in organic matter.[132–136] Although notable for the large size and great ages of the oldest trees, particularly in protected coves, many of these forests were not undisturbed. Native Americans managed some of the forest, often starting fires to clear land for agriculture

or to improve game habitat.[130] Hurricanes resulted in massive blowdowns, particularly along mountain ridges.[133]

Agriculture
European settlement sharply altered the landscape of the eastern forest in new ways. As settlers moved from the Atlantic coast toward the Mississippi River between the years of 1620 and 1872, approximately half of the eastern deciduous forest was cleared.[137] Effects of forest clearing on the abiotic and biotic factors of a forest were long lasting,[138–140] and recovery rate was variable.[139–141]

Much of the eastern deciduous forest was converted to agricultural use,[138,140] a trend that peaked in the mid 1800s.[142] On steep terrain throughout Appalachia, marginal farmlands were abandoned more frequently than they were cleared in the 1900s, a pattern that continues today. From 1973 to 2000, in the eastern United States, there was a net gain in forest cover.[143]

Despite the positive trend of forest cover increase in the past several decades, postagricultural recovery of native forest diversity has been slow.[144] As an obligate understory species with a short-lived seed bank, ginseng was vulnerable to extermination in areas where forest was converted to agricultural land. There are conflicting conclusions regarding the effects of prior agriculture land-use on current forests; most evidence, however, suggests that postagricultural secondary forests exhibit lower herbaceous understory biodiversity[139] and altered species composition.[145] Lower levels of diversity are the result of many factors, including colonization limitations[144,145] and residual environmental effects.[146–148] Given ginseng's large seeds, dominant mode of dispersal by gravity, and low seed numbers, propagule limitation could play an important role in limiting the rate of repopulation of postagricultural forests by ginseng.[149]

Timbering
In the mid-1800s, large-scale logging operations developed east of the Mississippi River in order to meet demand for wood fuel and wood products.[130] From the mid-1800s to the early 1900s, there was an eightfold increase in the rate of lumber production nationwide.[130] In addition to the disturbances caused by the removal of large amounts of timber, the slash left behind by timbering operations was often ignited by sparks emitted by steam-powered railcars

that carried timber to and from the sawmills.[150] By the 1920s, much of the merchantable timber in the eastern deciduous forest was gone and the number of active timbering operations began to decline.[131]

Second growth timber in the same forests that were exploited in the early 1900s have since reached merchantable size.[151] Timber market models predict that timber harvests nationwide will increase by one-third between 1995 and 2040.[152] The cyclical disturbance typical of timber rotations will likely alter, and possibly degrade, a wide range of ginseng habitats. As understory environments change, the future of ginseng in these forests is uncertain.

The extent of habitat change caused by timber harvest is not fixed; the changes that occur depend upon the type and the intensity of timber harvest.[153] Following timbering, changes in understory microclimates occur. Light levels, mean temperature, and temperature fluctuations near the herbaceous layer increase, while humidity and soil moisture decline.[153–155] A decrease in total canopy cover resulting from timber harvest alters the interaction between the understory and higher strata and alters competition within the herbaceous layer.[153] Moreover, the susceptibility of the forest area to invasion by nonnative species increases following a timbering event.[156,157]

A consensus regarding the response of the herbaceous understory to a wide range of timber harvest disturbance gradients has not been reached.[1,158,159] In addition, the response of ginseng populations to the environmental changes caused by timbering has not been investigated, but will likely depend on the disturbance intensity of each separate timbering event. Anecdotal information suggests that ginseng may be preadapted to sporadic, intermediate intensity canopy-opening events that are common in mature forests. For example, enhanced growth of ginseng plants has been observed one and two years following a single tree-fall event. Conversely, in areas where large amounts of the canopy were damaged in an ice storm, ginseng plants became yellow and dried before the end of the growing season. Interestingly, ginseng harvesters have reported the emergence of large ginseng plants in the early years following clear cutting; however, no populations have been followed through and beyond a timbering event to determine the net long-term effect.[12]

Wild ginseng populations are not confined to the small percentage of undisturbed old growth forest remaining in the United States; therefore, populations must have persisted following the extensive timbering that took place in the late 1800s and early 1900s. As an ongoing source of cyclical disturbance to the understory, the extent to which different timber harvest practices affect the population dynamics of ginseng must be quantified in order to place this widespread source of disturbance into broad perspective.

Surface mining

The natural range of ginseng overlaps that of the rich coal deposits found in the Anthracite, Appalachian, Eastern Interior, and Western Interior coal regions of the United States.[19,160] Coal mining became a boom industry at the beginning of the twentieth century.[161] As a result of improved technology, ease of extraction, and increased demand for coal, surface mining gained in popularity and scale in the 1950s.[162] The controversial method of surface mining known as mountain top removal (MTR) mining has been used since the 1970s,[163] and expanded in the 1990s[164] due to the demand for low-sulfur bituminous coal in eastern Kentucky and southern West Virginia.[164,165]

Before 1977, surface-mined areas were left unreclaimed about 40% of the time.[166] In 1977, the U.S. Government adopted the Surface Mining Control and Reclamation Act (SMCRA) in an effort to ameliorate the environmental damage imposed by mining. SMCRA requires that mined lands be reclaimed to the approximate original contour; the process and type of reclamation, however, is determined by the mine operator.[167] Often the postmining landscape is planted with grasses[168] because this type of reclamation is typically cheaper and faster.[169]

Most coal extraction in the eastern United States now occurs *via* surface mining rather than underground mining.[170] In six of the northern and central Appalachian states, about 1.1 million hectares of forest have been directly affected by surface mining.[171] According to estimates by the E.P.A., 330,225 hectares of forest in southern Appalachia were destroyed by MTR mining between 1992 and 2012.[172] There are also indirect effects from surface mining, such as changes in soil chemistry,[171] soil fertility,[173] water quality,[174] nutrient cycling,[175] and increased flooding.[176,177] Surface mining increases habitat fragmentation[178] and reduces the amount of interior forest in adjacent communities.[172] While succession on previously un-reclaimed surface mines shows tree recruitment,[179] the native herbaceous understory is often absent or sparse.[139] Current trends in reclamation include restoring the surface-mined land using the Forestry Reclamation Approach (FRA). The FRA's primary goal is to increase timber production on previously surfaced-mined lands, with the belief that natural succession will generate a suitable habitat for herbaceous species over time.[168] FRA is a recent development, and it is early to judge whether overall forest diversity restoration is accelerated by this approach.

The direct and indirect loss of the eastern deciduous forest habitat caused by surface mining has no doubt extirpated thousands of ginseng populations. Based on extensive quadrat censusing, an estimate was made by McGraw *et al.*[21] that there is a ginseng density of 18.26 plants per hectare in deciduous forests of the region, including areas near surface mining. Assuming this density is a realistic estimate, in the past 20 years about six million ginseng plants have been lost due to MTR mining alone. The ability of ginseng to grow on old unreclaimed sites and mined lands that have been reclaimed to a forest is not known. As with postagricultural lands, restoration of ginseng to these sites would likely require assisted relocation, or it would require decades if not centuries to occur naturally.

Acid deposition

The effects of acid deposition on terrestrial and aquatic ecosystems have become global concerns over the past several decades.[180–182] Sulfur dioxide and nitrogen oxides are commonly released into the atmosphere as a result of automobile exhaust and industrial plants, among other sources, falling as acid precipitation on vast areas of forests downwind.[180,183,184] Acid precipitation alters soil properties, and particularly affects the cation balance in the soil, resulting in depletion of calcium and other important cations.[180,184] Reduced available calcium and lowered pH in forest soils inhibits growth in many plant species.[184–187] Research suggests that ginseng populations located in soils rich in calcium typically grow larger and are less susceptible to disease than ginseng grown in less calcium-rich soils.[188,189] In fact, in southern Appalachia,

ginseng grown on calcium-poor sites actually displayed stunted growth.[188,189]

No studies have directly tested the effect of acid precipitation on natural populations of ginseng; however, much work has focused on effects on tree species. The high-elevation mountains of the eastern United States are of particular interest in the study of acid precipitation, as the thin soils are subjected to high amounts of acid precipitation in the form of fog as well as rain and snow. A decline in the growth of tree species such as *Picea rubens* and *A. saccharum* has been detected in many high elevation study sites in the northeast United States.[184–187] Far less research has been performed to address the response of the herbaceous understory to acid precipitation. However, Lodhi[190] found that experimental acid rain applications caused a significant reduction in the biomass of multiple herbaceous species found in Missouri.

Suburban sprawl

The rate of suburbanization has increased substantially in the eastern United States since 1973. Over a 27-year period, 1.9 million hectares have been converted from forest to suburban landscapes.[143] Theobald[191] predicted that by 2020, urban and suburban development will increase across the United States by 2.2% and 14.3% respectively. As urban and suburban areas spread, forest land area is both lost and fragmented.[192] Habitat fragmentation results in decreased species richness and abundance.[193] Furthermore, fragmentation reduces pollinator abundance and diversity, which may lead to reduced seed set among the plants in the fragmented populations.[194] The division of populations into smaller subunits also increases the susceptibility of ginseng to population-level extinction caused by demographic and environmental stochasticity.[195] Increase in the frequency of edge environments with fragmentation can, in turn, affect the viability of forest interior species[196] such as ginseng.

Synthesis

Figure 1 of this review represented a simple starting point for studying factors affecting herbaceous plants in the eastern deciduous forest, using ginseng as a phytometer. By emphasizing individual plant responses to environmental factors, then linking them to demographic effects, the body of research

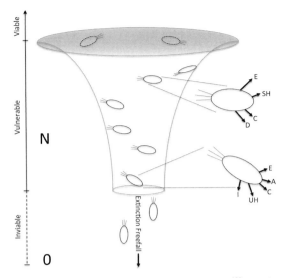

Figure 9. Representation of the extinction vortex[198] containing ginseng populations subject to negative and positive forces acting differentially in each population. Vectors of change: E = exponential growth, SH = stewardship harvest, C = climate warming, D = deer browse, A = acidification, UH = unsustainable harvest, I = inbreeding. N is the population size.

to date suggests a more nuanced understanding of the present status of ginseng populations (Fig. 9).

Thousands of small natural populations of ginseng exist in habitats experiencing a unique set of negative and positive forces, resulting in population decline (down the vortex), stability, or increase (up the vortex). Different forces will act with changing probabilities as a function of position within the vortex. The reviewed studies show three dominant forces acting at all population sizes; harvest, deer browse, and climate change. All three appear to be exerting downward pressure regardless of population size. Toward the bottom of the vortex, at the lowest N's, additional factors such as the Allee effect and inbreeding depression are expected to be more significant, accelerating decline. The proportions of populations experiencing each trajectory are unknown. The precise boundaries between inviable and vulnerable populations and between vulnerable and viable populations are not known. The categories represent a theoretical construct, and vulnerability to extinction is a continuous probability function of population size, but we do not yet have a thorough understanding of this function.

What stands out as we survey vectors affecting ginseng population fates is the pervasiveness of

direct and indirect human actions. Massive landscape level changes (agriculture, timbering) since European settlement had direct negative effects on forest herb communities. Layered on top of this in the case of ginseng, the long history of harvest was responsible for reducing population sizes to the point where most are now small, and likely fall into the "vulnerable" or "inviable" class. Species with life histories such as ginseng may recover from such effects, albeit slowly. More subtle but widespread environmental changes such as acid precipitation or ozone pollution effects are also, obviously, caused by humans, though their effects on ginseng are unknown. More recent human effects, such as surface mining and suburban sprawl, eliminate habitat, shrinking the funnel and reducing the total number of populations substantially. These latter changes are directional, ongoing, and likely permanent.

Two of the indirect effects of human actions appear overwhelmingly important to the future of species such as ginseng. Deer overpopulation and consequent overbrowsing is an indirect effect of mismanagement of the ecosystem, in turn caused by very effective management of the deer herd for the hunting constituency.[94] Climate change is an indirect effect of human reliance on fossil fuel burning and, thus far, a lack of political will to solve the problem.[199] Large uncertainties exist for both factors. Overbrowsing by deer will invariably have negative effects on ginseng populations as long as deer populations are high; however, the future of high deer populations is not guaranteed: ecological surprises, such as uncontrolled disease outbreaks could reduce the herd. Alternatively, society may reach a tipping point in public opinion, and the desire for deer herd control could outweigh the political sway currently held by hunters, prompting changes in the way deer are managed.

While there is uncertainty about whether the political will can be summoned to address climate change, the scientific consensus is that, even if this occurs, major climate shifts are already happening and will continue. More uncertainty surrounds the complex biotic response to climate change, and the consequences of those responses for whether a particular native species will persist or go extinct in a changing world.[128] For ginseng, the response of fungal diseases or insect herbivores to a warmer, possibly wetter, climate may be more important than direct effects on growth. While invasive plant species are not presently having large, pervasive effects on ginseng, this could change rapidly as native plants decline in forest interiors and niche space is opened for effective colonizers. These ecological surprises may trigger future changes in prevalent vectors acting on each population in ways we cannot at present predict.

If species such as ginseng are to persist, powerful natural forces may play roles. Every species has the potential to grow exponentially, even those with slow life histories such as ginseng. The fact that ginseng exists at historically low population sizes suggests that it may be far below its carrying capacity, and therefore released to a degree from density-dependent population regulation. In addition, we have very little understanding of ginseng metapopulation dynamics: in particular, how frequently can long-distance dispersal occur and how often are new populations successfully founded? We know natural selection may bring about fundamental changes in the ability of populations to resist negative factors such as deer browse and climate change, and to take advantage of opportunities such as invading postagricultural forests. Research is urgently needed to understand whether sufficient additive genetic variation exists within populations to allow adaptation to play a significant role in persistence through environmental change.

Owing to the pace and intensity of environmental change brought about by direct and indirect effects of human actions, natural processes may be inadequate to prevent a rapid flushing of populations down the extinction vortex. In this case, for the species to survive, human intervention may be required. Such intervention can take many forms, and successful intervention will require more fundamental scientific understanding. For example, explicit efforts can be made to reintroduce or restore ginseng populations to formerly occupied habitats. The best strategies for accomplishing this, however, have not been determined and will rely on improved predictive understanding of soil requirements as well as the importance of the genetic background of seed sources.

As climate change continues, managed relocation of populations may be necessary to match ecotypes with their climatic requirements. This will be challenging without a better understanding of the indirect effects of climatic adaptation, and the ways in which populations are adapted to local sites independent of climate. Questions concerning the scale of relocations necessary for success, and the consequences of the introduction of new genotypes for extant populations, need to be addressed.

While perhaps unique to wild harvested species, the opportunity for altering population fates by improving management strategies exists, given our current understanding. Current ginseng harvest practices range from unsustainable, which can cause rapid population decline, to stewardship, which may grow populations. Unethical behavior by harvesters is partly to blame for the former; clearly law enforcement is presently inadequate to stem such behavior. An obvious solution is to house ginseng management programs within state agencies, such as Wildlife Departments, that have natural resource law enforcement as part of their mission.[63] In addition, harvest regulations have been slow to change in response to better ecological understanding: while harvest seasons have evolved, minimum age requirements are still national policy even though size is clearly a far better predictor of reproductive success than age. Given that harvesters can be stewards if they plant adequate numbers of mature seeds, encouraging this behavior with size-minimums and optimized harvest seasons can change harvest from a downward vector to a neutral or upward vector for population change. Replacing the age requirement would also allow planting of detached rhizomes as a means of clonally propagating the harvested individual and further mitigating harvest effects (presently, intact rhizomes are required to prove that the age-requirement is met).

If both natural and human-initiated processes do not reverse the loss of populations from the extinction vortex, increased rarity will have one further consequence as long as the wild ginseng market persists: prices for roots would rise rapidly. This would trigger further unethical harvest, followed by harvest ban, and the inevitable black market for wild ginseng roots. At that point, conservation of the species would become increasingly difficult, with challenges similar to those being carried out on behalf of the tiger, rhinoceros or elephant. The difference in the case of tigers, rhinoceroses, or elephants is that they are large, charismatic mammals that have developed a constituency all their own. For a modest herbaceous plant such as American ginseng, a more likely fate would be that experienced by its sister species, *P. ginseng*, on the Asian continent, long ago. The species would persist in cultivation, but it would continue its evolution in that environment into a different organism, eventually losing traits that would allow persistence in the wild. The species, *P quinquefolius*, as we know it, would be extinct.

Although the scenarios depicted by our current understanding of ginseng ecology and the effects of environmental change are discouraging, the scientific research accomplished so far has been valuable in clarifying areas of real concern (unsustainable harvest practices, overbrowsing by deer, and climate change) and pointing to possible solutions (improved management policies, rebalancing the community structure, and managed relocation). As a plant species with medicinal value in Asian culture, economic and cultural value in Appalachia, value as part of a functional, diverse forest understory, and potential value for western medicine, ginseng's worth as a target for conservation is evident. To the extent that ginseng is a phytometer for general effects on herbaceous plants, population studies of ginseng illustrate the magnitude of the diverse challenges faced by plants in a changing world.

Conflicts of interest

The authors declare no conflicts of interest.

Supporting Information

Additional Supporting Information may be found in the online version of this paper.

Fig. S1. Artist's rendering of a small cluster of American ginseng plants in autumn in a natural understory setting. Watercolor by Susan Bull Riley.

References

1. Gilliam, F.S. 2007. The ecological significance of the herbaceous layer in temperate forest ecosystems. *Bioscience* **57:** 845–858.
2. Schimel, D., W. Hargrove, F. Hoffman & J. MacMahon. 2007. NEON. A hierarchically designed national ecological network. *Front. Ecol. Env.* **5:** 59.
3. Clements, F.E. & G.W. Goldsmith. 1924. *The Phytometer Method in Ecology.* Carnegie Institution of Washington Publication No. 356.
4. Antonovics, J. & R.B. Primack. 1982. Experimental ecological genetics in Plantago: VI. The demography of seedling transplants of *P. lanceolata. J. Ecol.* **70:** 55–75.
5. Antonovics, J.K. Clay & J. Schmitt. 1987. The measurement of small-scale environmental heterogeneity using clonal transplants of *Anthoxanthum odoratum* and *Danthonia spicata. Oecologia* **71:** 601–607.
6. Billings, W.D. 1964. *Plants, Man and The Ecosystem.* Wadsworth. Belmont, CA.
7. Carpenter, S.G. & G. Cottam. 1982. Growth and Reproduction of American Ginseng (*Panax quinquefolius*) in Wisconsin, USA. *Can. J. Bot.* **60:** 2692–2696.
8. Lewis, W.H. & V.E. Zenger. 1982. Population dynamics of the American ginseng, *Panax quinquefolium* (Araliaceae). *Am. J. Bot.* **69:** 1483–1490.
9. Lewis, W.H. 1988. Regrowth of decimated population of *Panax quinquefolium* in a Missouri climax forest. *Rhodora* **90:** 1–5.
10. Charron, D. & D. Gagnon. 1991. The demography of northern populations of *Panax quinquefolium* (American ginseng). *J. Ecol.* **79:** 431–445.
11. Nantel, P., D. Gagnon & A. Nault. 1996. Population viability analysis of American ginseng and wild leek harvested in stochastic environments. *Cons. Biol.* **10:** 608–621.
12. Taylor, D.A. 2006. *Ginseng, the Divine Root: The Curious History of the Plant that Captivated the World.* Algonquin Books of Chapel Hill. Chapel Hill.
13. Johannsen, K. 2006. *Ginseng Dreams. The Secret World of America's Most Valuable Plant.* The University Press of Kentucky. Lexington. 215 pp.
14. Pritts, K.D. 1995. *Ginseng: How to Find, Grow and Use America's Forest Gold.* Stackpole. Mechanicsburg, PA. 150 pp.
15. McGraw, J.B. 2001. Evidence for decline in stature of American ginseng plants from herbarium specimens. *Biol. Cons.* **98:** 25–32.
16. Robbins, C.S. 2000. Comparative analysis of management regimes and medicinal plant trade monitoring mechanisms for American ginseng and goldenseal. *Conserv. Biol.* **14:** 1422–1434.
17. McGraw, J.B., M.A. Furedi, K. Maiers, *et al.* 2005. Berry ripening and harvest season in wild American ginseng. *Northeast. Natur.* **12:** 141–152.
18. USFWS. 2011. Advice for the export of roots of wild and wild-simulated American ginseng (*Panax quinquefolius*) lawfully harvested during the 2011 harvest season in 19 States. Department of the Interior, United States Fish and Wildlife Service Division of Scientific Authority.
19. USDA, NRCS. The PLANTS Database. 2012. Retrieved August 22, 2012, from http://plants.usda.gov/java/ profile?symbol=paqu
20. Anderson, R.C., J.S. Fralish, J.E. Armstrong & P.K. Benjamin. 1993. The ecology and biology of *Panax quinquefolium* L. (Araliaceae) in Illinois. *Am. Mid. Nat.* **129:** 357–372.
21. McGraw, J.B., S.M. Sanders & M.E. Van der Voort. 2003. Distribution and Abundance of *Hydrastis canadensis* L. (Ranunculaceae) and *Panax quinquefolius* L. (Araliaceae) in the Central Appalachian Region. *J. Torr. Bot. Club* **130:** 62–69.
22. Fountain, M.S. 1982. Site factors associated with natural populations of ginseng in Arkansas. *Castanea* **47:** 261–265.
23. Albrecht, M.A. & B.C. McCarthy. 2009. Seedling establishment shapes the distribution of shade-adapted forest herbs

across a topographical moisture gradient. *J. Ecol.* **97:** 1037–1049.

24. Rabinowitz, D. 1981. Seven forms of rarity. In *The Biological Aspects of Rare Plant Conservation.* H. Synge, Ed. Wiley. New York.

25. Cruse-Sanders, J.M. & J.L. Hamrick. 2004a. Spatial and genetic structure within populations of wild American ginseng (*Panax quinquefolius* L., Araliaceae). *J. Hered.* **95:** 309–321.

26. Mooney, E. & J. McGraw. 2007a. Alteration of selection regime resulting from harvest of American ginseng, *Panax quinquefolius*. *Conserv. Gen.* **8:** 57–67.

27. McGraw, J.B., S. Souther & A.E. Lubbers. 2010. Rates of harvest compliance with regulations in natural populations of American ginseng (*Panax quinquefolius* L.). *Nat. Areas J.* **30:** 202–210.

28. Lewis, W.H. & V.E. Zenger. 1983. Breeding systems and fecundity in the American ginseng, *Panax quinquefolium* (Araliaceae). *Am. J. Bot.* **70:** 466–468.

29. Mooney, E.H. & J.B. McGraw. 2009. Relationship between age, size and reproduction in populations of American ginseng, *Panax quinquefolius* (Araliaceae), across a range of harvest pressures. *Ecoscience* **16:** 84–94.

30. Mahall, B.E. & F.H. Bormann. 1978. A quantitative description of the vegetative phenology of herbs in a northern hardwood forest. *Bot. Gaz.* **139:** 467–481.

31. Schlessman, M.A. 1985. Floral biology of American ginseng (*Panax quinquefolium*). *Bull. Torr. Bot. Club* **112:** 129–133.

32. Fiebig, A.E., J.T.A. Proctor, U. Posluszny & D.P. Murr. 2001. The North American ginseng inflorescence: development, floret abscission zone, and the effect of ethylene. *Can. J. Bot.* **79:** 1048–1056.

33. Proctor, J.T.A., M. Dorais, H. Bleiholder, *et al.* 2003. Phenological growth stages of North American ginseng (*Panax quinquefolius*). *Ann. Appl. Biol.* **143:** 311–317.

34. Stoltz, L.P. & J.C. Snyder. 1985. Embryo growth and germination of American ginseng seed in response to stratification temperatures. *Hort. Science* **20:** 261–262.

35. Baskin C.C. and J.M. Baskin. 1998. *Seeds. Ecology, Biogeography, and Evolution of Dormancy and Germination.* Academic Press. New York. 666 pp.

36. Proctor, J.T.A. & D. Louttit. 1995. Stratification of American ginseng seed: embryo growth and temperature. *Korean J. Ginseng. Sci.* **19:** 171–174.

37. Stoltz, L.P. & P. Garland. 1980. Embryo development of ginseng seed at various stratification temperatures. *Proceedings of the Second National Ginseng Conference.* Missouri Department of Conservation, Jefferson City, MO.

38. Mooney, E.H. & J.B. McGraw. 2007b. Effects of self-pollination and outcrossing with cultivated plants in small natural populations of American ginseng, *Panax quinquefolius* (Araliaceae). *Am. J. Bot.* **94:** 1677–1687.

39. Souther, S. & J.B. McGraw. 2011. Evidence of local adaptation in the demographic response of American ginseng to interannual temperature variation. *Conserv. Biol.* **25:** 922–931.

40. McGraw, J.B., M.A. Furedi & E. Mooney. 1986. Environmental effects on seed bank dynamics of American ginseng. Annual Meeting of the Ecological Society of America Abstracts: http://abstracts.co.allenpress.com/pweb/esa2006/document/60671.

41. Bierzychudek, P. 1982. Life histories and demography of shade-tolerant temperate forest herbs: a review. *N. Phytol.* **90:** 757–776.

42. Van der Voort, M. 2005. An ecological study of *Panax quinquefolius* in central Appalachia: Seedling growth, harvest impacts and geographic variation in demography. Doctor of Philosophy in Biology, West Virginia University, Morgantown, WV.

43. Fournier, A.R., A. Gosselin, J.T.A. Proctor, *et al.* 2004. Relationship between understory light and growth of forest-grown American ginseng (*Panax quinquefolius* L.). *J. Am. Soc. Hort. Sci.* **129:** 425–432.

44. Hull, J.C. 2002. Photosynthetic induction dynamics to sunflecks of four deciduous forest understory herbs with different phenologies. *Int. J. Pl. Sci.* **163:** 913–924.

45. Kudo, G., T.Y. Ida & T. Tani. 2008. Linkages between phenology, pollination, photosynthesis, and reproduction in deciduous forest understory plants. *Ecology* **89:** 321–331.

46. Jochum, G.M., K.W. Mudge & R.B. Thomas. 2007. Elevated temperatures increase leaf senescence and root secondary metabolite concentrations in the understory herb *Panax quinquefolius* (Araliaceae). *Am. J. Bot.* **94:** 819–826.

47. Schluter, C. & Z.K. Punja. 2000. Floral biology and seed production in cultivated North American ginseng (*Panax quinquefolius*). *J. Am. Soc. Hort. Sci.* **125:** 567–575.

48. Schlessman, M.A. 1987. Gender modification in North American ginsengs. *BioScience* **37:** 469–475.

49. Duke, J.A. 1980. Pollinators of *Panax*? *Castanea* **45:** 141.

50. Catling, P.M. & K.W. Spicer. 1995. Notes on economic plants (Araliaceae). *Econ. Bot.* **49:** 99–102.

51. Van der Voort, M.E., B. Bailey, D.E. Samuel & J.B. McGraw. 2003. Recovery of populations of goldenseal (*Hydrastis canadensis* L.) and American ginseng (*Panax quinquefolius* L.) following harvest. *Am. Mid. Nat.* **149:** 282–292.

52. Schluter, C. & Z.K. Punja. 2002. Genetic diversity among natural and cultivated field populations and seed lots of American ginseng (*Panax quinquefolius* L.) in Canada. *Int. J. Plant. Sci.* **163:** 427–439.

53. Cruse-Sanders, J.M. & J.L. Hamrick. 2004b. Genetic diversity in harvested and protected populations of wild American ginseng, *Panax quinquefolius* L. (Araliaceae). *Am. J. Bot.* **91:** 540–548.

54. Hackney, E.E. & J.B. McGraw. 2001. Experimental demonstration of an Allee effect in American ginseng. *Conserv. Biol.* **15:** 129–136.

55. Caswell, H. 2001. *Matrix Population Models.* Sinauer. Sunderland, MA. 722 pp.

56. Van der Voort, M.E. & J.B. McGraw. 2006. Effects of harvester behavior on population growth rate affects sustainability of ginseng trade. *Biol. Cons.* **130:** 505–516.

57. Furedi, M.A. 2004. Effects of Herbivory by White-Tailed Deer (*Odocoileus virginianus* Zimm.) on the Population Biology of American Ginseng (*Panax quinquefolius* L.). Ph.D. dissertation. West Virginia University, Morgantown.

58. McGraw, J.B. & M.A. Furedi. 2005. Deer browsing and population viability of a forest understory plant. *Science* **307:** 920–922.

59. Farrington, S.J., R-M Muzika, D. Drees & T.M. Knight. 2008. Interactive effects of harvest and deer herbivory on the population dynamics of American ginseng. *Cons. Biol.* **23**: 719–728.

60. Morris W.F. & D.F. Doak. 2002. *Quantitative Conservation Biology. Theory and Practice of Population Viability Analysis.* Sinauer. Sunderland, MA. 480 pp.

61. Souther, S. 2011. Demographic Response of American Ginseng (*Panax quinquefolius* L.) to Climate Change. Doctor of Philosophy, West Virginia University, Morgantown, WV.

62. Hardin, G. 1968. The tragedy of the commons. *Science* **162**: 1243–1248.

63. Bailey, B. 1999. Social and Economic Impacts of Wild Harvested Products. Ph.D. Dissertation. Morgantown, West Virginia.

64. Freese, C.H. 1998. *Wild Species as Commodities.* Island Press. Washington, DC.

65. Buchert, G.P., O.P. Rajora, J.V. Hood & B.P. Dancik. 1997. Effects of harvesting on genetic diversity in old-growth eastern white pine in Ontario, Canada. *Conserv. Biol.* **11**: 747–758.

66. Sebbenn, A.M., B. Degen, V.N.C.R. Azevedo, *et al.* 2008. Modelling the long-term impacts of selective logging on genetic diversity and demographic structure of four tropical tree species in the Amazon forest. *For. Ecol. Manage.* **254**: 335–349.

67. Allendorf, F.W. & J.J. Hard. 2009. Human-induced evolution caused by unnatural selection through harvest of wild animals. *Proc. Natl. Acad. Sci.* **106**: 9987–9994.

68. Darimont, C.T., S.M. Carlson, M.T. Kinnison, *et al.* 2009. Human predators outpace other agents of trait change in the wild. *Proc. Natl. Acad. Sci. USA* **106**: 952–954.

69. Edmands, S. 2007. Between a rock and a hard place: evaluating the relative risks of inbreeding and outbreeding for conservation and management. *Mol. Ecol.* **16**: 463–475.

70. Chung, M.Y. & J.D. Nason. 2007. Spatial demographic and genetic consequences of harvesting within populations of the terrestrial orchid *Cymbidium goeringii. Biol. Conserv.* **137**: 125–137.

71. Boehm, C.L., H.C. Harrison, G. Jung & J. Nienhuis. 1999. Organization of American and Asian ginseng germplasm using randomly amplified polymorphic DNA (RAPD) markers. *J. Am. Soc. Hort. Sci.* **124**: 252–256.

72. Grubbs, H.J. & M.A. Case. 2004. Allozyme variation in American ginseng (*Panax quinquefolius* L.): variation, breeding system, and implications for current conservation practice. *Conserv. Genet.* **5**: 13–23.

73. Lim, W., K.W. Mudge & L.A. Weston. 2007. Utilization of RAPD markers to assess genetic diversity of wild populations of North American ginseng (*Panax quinquefolium*). *Planta Med.* **73**: 71–76.

74. Schlag, E. & M. McIntosh. 2012. RAPD-based assessment of genetic relationships among and within American ginseng (*Panax quinquefolius* L.) populations and their implications for a future conservation strategy. *Genet. Resour. Crop. Evol.* **59**: 1553–1568.

75. Young, J., M. Eackles, M. Springmann & T. King. 2012. Development of tri- and tetra- nucleotide polysomic microsatellite markers for characterization of American ginseng (*Panax quinquefolius* L.) genetic diversity and population structuring. *Conserv. Genet. Res.* **4**: 833–836.

76. Loveless M.D. and J.L. Hamrick. 1984. Ecological determinants of genetic structure in plant populations. *Ann. Rev. Ecol. Syst.* **15**: 65–95.

77. Hamrick J.L. & M.J. W. Godt. 1996. Effects of life history traits on genetic diversity in plant species. *Phil. Trans. Roy. Soc. B.* **351**: 1291–1298.

78. Cruse-Sanders, J., J.L. Hamrick & J. Ahumada. 2005. Consequences of harvesting for genetic diversity in American ginseng (*Panax quinquefolius* L.): a simulation study. *Biodiv. Conserv.* **14**: 493–504.

79. Obae, S.G. & T.P. West. 2011. Effects of anthropogenic activities on genetic diversity and population structure of American ginseng (*Panax quinquefolius* L.) growing in West Virginia. *J. Hort. For.* **3**: 270–281.

80. Ellstrand, N.C. & D.R. Elam. 1993. Population genetic consequences of small population size: implications for plant conservation. *Ann. Rev. Ecol. Syst.* **24**: 217–242.

81. Byers, D.L. & D.M. Waller. 1999. Do plant populations purge their genetic load? Effects of population size and mating history on inbreeding depression. *Ann. Rev. Ecol. Syst.* **30**: 479–513.

82. Husband, B.C. & D.W. Schemske. 1996. Evolution of the magnitude and timing of inbreeding depression in plants. *Evolution* **50**: 54–70.

83. Williams, L. 1957. Ginseng. *Econ. Bot.* **11**: 344–358.

84. Carlson, A. 1986. Ginseng: America's botanical drug connection to the orient. *Econ. Bot.* **40**: 233–249.

85. Case, M.A., K.M. Flinn, J. Jancaitis, A. Alley & A. Paxton. 2007. Declining abundance of American ginseng (*Panax quinquefolius* L.) documented by herbarium specimens. *Biol. Conserv.* **134**: 22–30.

86. Law, W. & J. Salick. 2005. Human-induced dwarfing of Himalayan snow lotus, *Saussurea laniceps* (Asteraceae). *Proc. Natl. Acad. Sci. USA* **102**: 10218–10220.

87. McCabe, R.E. & T.R. McCabe. 1997. Recounting whitetails past. In *The Science of Overabundance: Deer Ecology and Population Management.* H.B. Underwood & J.H. Rappole, Eds.: 11–26. Smithsonian Institution Press, Washington, DC.

88. Alverson, W.S., D.M. Waller & S.L. Solheim. 1988. Forest too edge: edge effects in northern Wisconsin. *Conserv. Biol.* **2**: 348–358.

89. Van Deelen, T.R., K.S. Pregitzer & J.B. Haufler. 1996. A comparison of presettlement and present-day forests in two northern Michigan deer yards. *Am. Midl. Nat.* **135**: 181–194.

90. Waller, D.M. & W.S. Alverson. 1997. The white-tailed deer: a keystone herbivore. *Wildlife Soc. Bull.* **25**: 217–226.

91. Rooney, T.P. 2001. Deer impacts on forest ecosystems: a North American perspective. *Forestry* **74**: 201–208.

92. Gill, R.M.A. & V. Beardall. 2001. The impact of deer of woodlands: the effect of browsing and seed dispersal on vegetation structure and composition. *Forestry* **74**: 209–218.

93. Rooney, T.P. & D.M. Waller. 2003. Direct and indirect effects of white-tailed deer in forest ecosystems. *For. Ecol. Manag.* **181**: 165–176.

itself: the capacity of a system to accommodate a range of disturbances or state changes while retaining essential (or desired) structure and function.

Here, we propose a refinement of the HRV concept, drawing from decision-making frameworks across fire-prone parts of the world outside the United States. In particular, we borrow from an approach that emphasizes threshold-based dynamics to avoid key thresholds of potential concern (TPC), beyond which more drastic actions would have to be initiated to maintain desired system functioning.[18] We propose a *bounded ranges of variation* (BRV) framework, which accommodates socio-ecological thresholds and the need for landscape heterogeneity to increase future resilience. The BRV framework retains a basis in HRV, given that there is still much to learn about interactions between past fire regimes and ecosystem resilience.[19–21] We finish by describing how BRV may lend itself to different ecosystem management priorities, including carbon sequestration and controlling invasive species.

Historical range of variability, resilience, and thresholds

Whether baseline conditions for characterizing HRV reflect a reconstruction of past fire regimes (Fig. 1A) or past composition of vegetation mosaics (Fig. 1B), HRV typically aims to capture natural ecosystem functioning before modern human perturbations. For fire management, HRV would provide guidance on how to restore or emulate natural fire-related patterns in space and time. Note that Figure 1 depicts a more predictable (e.g., Gaussian, homogeneous) set of landscape dynamics than may have been typical in many areas.[16] It may also be difficult to establish meaningful and stable estimates of HRV from available records, given rates of ecosystem dynamics in relation to past climatic variation.[22] Regardless, an assumption in management using HRV is that ecosystems should be resilient to environmental fluctuations experienced over a reference period, or ecosystem integrity would not have persisted through these fluctuations.

In many ways, linking HRV and ecosystem function requires a somewhat Clementsian notion of ecosystem-as-organism, relying on a simplified view of ecosystem complexity and quasi-equilibrium states.[23] For example, evaluating vegetation patch composition with long-term landscape averages in mind[24, 25] is analogous to the shifting-mosaic steady

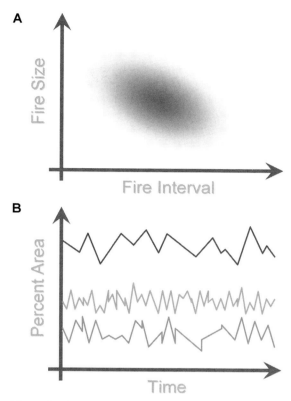

Figure 1. Alternative data sources for characterizing HRV. (A) A stylized range of past area burned and fire frequencies in an ecosystem. (B) A hypothetical time series of patch dynamics on a landscape, which contains three different (by color) vegetation types that fluctuate in how much of the landscape they occupy. Statistical measures of both central tendency and variation could be applied to describe HRV in either context.

state,[26] an early and equilibrium-oriented example of HRV. Although one chooses the spatial and temporal scales to characterize an ecosystem and its HRV, and this choice does not necessarily assume an equilibrium, the concept of resilience suggests the existence of a meta-stable state from which the ecosystem departs and to which it can return.[27] Whether an ecosystem exhibits dynamic equilibrium conditions will depend on the disturbance and recovery processes in question, but also the length of the historical reference period.[28] Given appropriate scales of space and time, it is nonetheless reasonable to relate HRV to some level of ecosystem persistence, stability, and resilience.

Ecological resilience is a measure of the capacity of a system to cope with disturbance and undergo change while retaining similar structure and function.[29–31] As resilience declines, it should take progressively milder disturbances to push the system into a different state (also termed *regime* or *basin of attraction*[32]). Here, the extremity of a disturbance, such as fire, is linked directly to HRV, as a potential deviation from natural fire patterns may cause transition to a different state. Because crossing this threshold can result in rapid and substantial changes in structure and function, one goal of management can be to increase resilience within particular basins of attraction, assuming the conditions defining basins are themselves somewhat stable (e.g., fire, herbivory, and rainfall causing regime shifts between vegetation states[33]).

Theoretically, these concepts assume expectations of equilibrium dynamics: within a state or regime the system tends toward a certain composition or landscape mosaic (e.g., a ball tending to settle at a bottom of a basin), similar to climatic-climax succession models. However, they also encompass an expectation of the ability of the system to adapt to a certain degree of variability (e.g., the ball can move within the bounds of the basin when pushed), complementary to the concept of HRV and the importance of landscape heterogeneity. Beyond those limits (the edge of the basin), the system moves toward a different eventual configuration (e.g., a ball passing into a new basin). The resulting differences in structure and function occur due to the development of different feedbacks among components of the system and with the biophysical environment. For example, some fire-prone systems naturally have alternative plant community states that tend to persist, such as shrubland and forest patches maintained by differing pyrogenic feedbacks on the same landscape.[34,35]

Measurement of ecological resilience is notoriously difficult,[36] in part because it depends on at least four factors: the existence of multiple states or basins of attraction, characterization of the different states, characterization of the deviation of a disturbance regime (trigger) from the natural disturbance regime, and the influence of additional factors that may change the shape of the basin of attraction. First, it should not be expected that all systems exhibit threshold dynamics; changes may be smoothly reversible within a single basin of attraction.[37] Concepts of resilience are therefore predominately applicable where there are alternate states. In these cases, management decisions must reflect the pos-

sibility of hysteretic effects (i.e., the effects of a past condition) that can make a shift from one to the other difficult or even impossible to reverse. Inside a single basin of attraction, recovery should be more straightforward (although it could still be slow) if the disturbance regime is returned to within HRV.

State variables are most often characterized by the dominance of different functional groups such as algal versus coral domination in coral reefs[38] and grass versus woody domination,[39] key species such as seagrass presence versus absence,[40] or key processes such as phosphorus to nitrogen limitation in lakes.[41] Landscape-scale dynamics, such as the HRV of different vegetation patches, are less commonly used state variables, but are certainly applicable to fire management.[24,25] All these measurements essentially describe a key system component, and they reflect the important feedbacks that change across thresholds. In some cases, crossing a threshold brings about a sudden, sharp, and dramatic change in the responding state variables, as in the shift from clear to turbid water in lake systems.[41] In other cases, although the dynamics and feedbacks of the system have flipped from one attractor to another, the transition in the state variables is more gradual, such as the change from a grassy to a shrub-dominated rangeland.[42] The speed of transition in state variables often reflects time lags associated with species life histories or legacies in ecosystem processes, which must be reflected in appropriate temporal scales used to characterize HRV.[22] One critical but often hard-to-test factor is that shifts in state variables persist for at least one complete turnover of the population.[43]

Meaningful measurements of ecosystem conditions must relate to changes in the key disturbance triggers that fundamentally alter important feedbacks. Here, such deviations constitute thresholds, which are sometimes implied in the application of HRV but seldom identified. For example, resilience is expected to depend on maintaining sources of regeneration or adaptation (see Box 1).[44] Maintaining seed sources and seed banks is thus critical for recovery and can affect the likelihood of thresholds being crossed. This aspect of resilience requires a landscape or regional perspective in management, because ensuring that species from adjacent areas can colonize is essential. Undesirable connectivity must also be considered, as it could allow the colonization of invasive or other species that may affect

Box 1. Diversity and resilience.

Functional diversity should relate to ecological resilience and disturbance HRV. Loss of a major functional group, such as apex predators, other consumers, or benthic filter-feeders, may cause drastic alterations in ecosystem functioning.[46–48] For example, overhunting and use of fire by humans some 30,000–40,000 years ago is thought to have caused a widespread and irreversible shift that resulted in an ecosystem of fire and fire-dominated plants.[49] Similarly, in systems that lack a specific functional group, the addition of just one species may dramatically change the structure and functioning of ecosystems.[39,50]

How species respond to disturbance (functional response) relative to how they affect function (functional effect) may be critical to ecosystem resilience;[46,51,52] this has been called response diversity[53] to environmental change among species that contribute to the same ecosystem function. By this logic, landscape and genetic diversity should also be important to resilience at some scale.[54]

While diversity in general can relate to resilience, some have distinguished between real and apparent redundancy.[30] More species do not lead to increased system performance where there is real functional redundancy (similar functional effect traits[55]). Furthermore, if this set of functionally redundant species does not exhibit any response diversity, they would not contribute to resilience. However, there has been little empirical work to determine whether these aspects of functional diversity can be

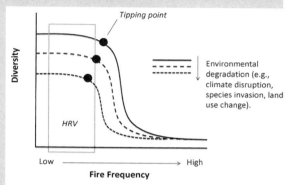

The upper response curve represents a functioning ecosystem state that is resilient (flat portion) to a range of fire frequencies within HRV; however, higher fire frequencies can lead to a threshold transition to a much less diverse ecosystem. Given a loss of diversity due to environmental degradation (shift to lower response curves), the ecosystem may be closer to a threshold transition point, even if fire frequencies are within HRV (lowest curve).

consistently determined a priori or how they relate to HRV; certainly, a conservative strategy would be to maintain high diversity at multiple levels, assuming that this taxonomic diversity will relate to functional diversity.

ecosystem function in undesirable ways, causing system reorganization and a breakdown in system-level processes following disturbance. Before system-level resilience can be understood, the boundary of the desirable basin of attraction must be defined, providing a metric against which the current system state is evaluated.

The capacity of a system to cope with a disturbance is related to more than just the component states and disturbance trigger. Other factors, such as climate and species pool limitations, can change the shape of the basin of attraction and thus affect adaptive capacity (i.e., the ability to avoid threshold transitions). Scale is an essential consideration,[45] as many important feedbacks occur across scales, such as dispersal connectivity and recovery. In ad-

dition, thresholds can be expected to occur at some scales but not others; for instance, recovery within a patch type might follow very different dynamics than patches across a landscape. The importance of landscape heterogeneity and its HRV emphasizes the interactions between these scales.

Challenges of implementing HRV

Once HRV has been characterized, it must be applied. Several examples are presented in recent reviews of HRV.[11,12] The most straightforward implementation has been to use a measure of deviation from the mean as a management guideline. While this may be helpful for comparing current and historical time series, a potentially risky assumption here is that distance from the mean is an isotropic risk in terms of ecological destabilization. In other words, avoiding both the left and right tails of the distribution in question (e.g., fires that are too small or too large) is implied as equally important, and this may not always be true.[56]

Other dangers in applying HRV include focusing only on one or a few parameters (e.g., fire frequency in Fig. 1A) or over-emphasizing measures of central tendency (e.g., mean patch composition in Fig. 1B). The U.S. LANDFIRE Program, for instance, used an average fire return interval from HRV to characterize how many multiples of this interval may have been missed in different ecosystems, to identify areas most in need of potential fuel treatments.[57] In contrast, some applications have gone to much greater lengths to characterize HRV of multiple patterns and processes for ecosystem management,[58,59] but this is not the norm. Regardless, some distillation of HRV complexity should be expected in its application to conservation and management problems. A key hurdle is interpreting the ecological meaning of the parameter space characterizing HRV and how it actually links to ecosystem resilience.

One of the largest threats to the relevance of HRV in ecosystem restoration and conservation is the possible emergence of no analogue or novel future conditions.[15,60–62] In such cases, past dynamics certainly could have less to offer in guiding future ecosystem management. It is important, however, to consider which pattern or process will potentially be unique. Are we concerned with climate norms, a top-down set of influences that are exogenous to a given ecosystem? Or are we concerned about soil moisture availability in an ecosystem, fine-scale patterns that can be constrained by abiotic characteristics of the landscape itself? Alternatively, the focus may be species assemblages, the patch mosaic of vegetation types, or fire regimes, all of which emerge as a result of cross-scale interactions and may therefore show some inertia in the face of climate change. Until there is compelling evidence for no analogue fire regimes in future climate scenarios, HRV for fire characteristics should remain relevant in ecosystem management (see Box 2).

It is worth noting that those advocating a move away from using HRV[15–17] also emphasize the heterogeneity of forest composition and pattern as a primary mechanism for facilitating resilience to climate change, arguing that such complexity was the norm for many prehistoric mixed conifer forests.

Therefore, relatively high landscape-scale heterogeneity is likely within the bounds of HRV and is a desirable characteristic, despite the difficulty it poses for specific management prescriptions. As opposed to disregarding HRV, a goal should be to develop methods and tools that more fully utilize HRV in landscape-scale planning and management.[63]

Specifying boundaries in HRV

Although the U.S. literature has addressed thresholds and ranges of variation in ecosystem dynamics,[13,37,42,66–69] to our knowledge, none have suggested the potential merger of HRV with a threshold-based framework. Conceptually, the closest to this merger may be the call for characterizing bounded variation[42] or the acceptable range of variation[66] in ecosystem dynamics. In contrast, some HRV-related concepts have been integrated in a TPC context,[70,71] although a lack of information about historical ecosystem dynamics has probably limited their application.

TPCs were originally envisioned as warning signs of unacceptable environmental change in national parks of South Africa, described as "upper and lower levels of change in selected biotic and abiotic variables."[18] These management thresholds are intended to precede and avoid ecological threshold responses. Because TPCs represent management decision points, fraught with imperfect information and evolving needs, the TPC framework has been refined to incorporate adaptive management.[72–75] In essence, the TPC approach reduces the probability of a system state change by anticipating the boundaries of the desired basin of attraction, including both physical and biological processes and feedbacks between the two.[37] Management triggers can incorporate more species- or process-specific information when knowledge of particular ecological thresholds is available. From this understanding comes a defined set of system parameters to assess the current state condition, relative to the desired distance from different thresholds.[18] Outside of South Africa, the TPC framework has been applied to ecosystem management in fire-prone parts of Australia.[76]

For fire-related conservation issues, concepts from TPC can inform HRV-based management in important ways. Incorporation of thresholds and basins of attraction in HRV, with the express goal

Box 2. No analogue fire activity in the future?

Because fire is affected by vegetation and productivity patterns that are partially constrained by topography, shifts in fire regimes may be less pronounced than those in coarse-scale climate parameters. The choice of downscaling approach is therefore important to realistically capture future climate-driven processes. The most common approach has been statistical downscaling, in which projections are interpolated and calibrated against local weather station data; the resulting spatial scale is often relatively coarse (e.g., tens of km). Process-based models can also be driven by coarse-scale climate variables, simulating fine-scale and topographically influenced micro-climate or water availability predictions.[55,64]

Future fire probabilities in California, for example, have been projected using fine-scaled eco-hydrology model outputs driven by different climate change scenarios.[65] Comparing a warmer–drier future with one representing a warmer–wetter future, the figure here shows that shifts in fire frequencies should be expected across nearly all elevations and time periods in a topographically complex region. Even so, most cases still show substantial overlap between current and future fire probabilities. Therefore, future no analogue fire frequencies could be relatively uncommon.

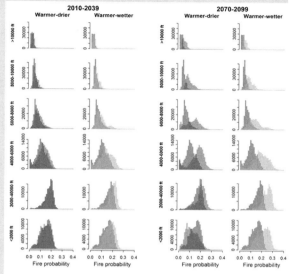

The area examined here covers several dozen vegetation types and life history strategies across ∼ 20,000 km² in the southern half of the Sierra Nevada Mountain ecoregion in California. The gray backdrop distribution of fire probabilities in each elevation class represents modeled baseline historical frequencies in the 1971–2000 period, and the colored distributions indicate shifts under future climate scenarios.

of managing resilient systems, can be considered bounds in our BRV approach. Rather than managing around observed system conditions, BRV accommodates the full range of possible conditions, recognizing the potential for a fire event that is in the unobserved tails of the distribution (e.g., very small or very large fires). Rare events such as large or very severe fires are often out of human control, yet some can facilitate landscape heterogeneity.[21] Therefore, while these events may be uncharacteristic and a significant departure from the mean, some may be acceptable or even desired if they do not approach a management boundary that precedes an ecological threshold.

Specifying boundaries for the application of BRV requires identification of the potential range of alternative states or losses (e.g., functionality or local extirpations), along with quantification of the

threshold level of disturbances causing undesirable changes. Such undesirable changes can also include human preferences and perceptions of what is unacceptable, resulting in systems of socio-ecological boundaries.[77] Although it is unlikely that precise disturbance characteristics associated with ecological threshold transitions will be identifiable in many circumstances, general steps for an iterative implementation of BRV are provided below. By integrating monitoring, science, and adaptive management to refine unacceptable thresholds under a changing environment,[37,73,74] BRV could provide a framework that accommodates heterogeneity and natural variation, where resilience is still the primary goal.

BRV: fire and nonnative plants

Nonnative plants, and their potential to transform ecosystems,[80] provide an example of a change agent that can push a system beyond an ecological threshold. The term *transformation* is used to imply a change in the dominant species present, the physical structure of the system, and the internal ecosystem processes resulting from and potentially reinforcing the compositional changes. Transformation is essentially passing beyond a TPC or out of a basin of attraction into a different and undesirable basin. The ability of the system to return to the prior basin and HRV is assumed to be low. Some of the most dramatic or rapidly occurring examples of transformation are alterations of fire regime caused by changes in fuel structure or distribution due to alien plant invasions.[39,81,82] Individual or multiple fires can trigger the transformation. Potential consequences for conservation include declines in native species cover and richness that are both short and long term,[83–86] loss of topsoil or carbon storage or other functions,[87,88] and loss of wildlife habitat.[89] Identifying boundaries and maintaining the system within BRV requires understanding the importance of environmental or disturbance regime thresholds to keep transforming invaders out of an ecosystem, as well as abundance thresholds of the invader beyond which the probability of ecosystem transformation is greatly enhanced. Thus the BRV for systems where invasive species are of concern should include both biotic (invader abundance) and abiotic (fire regime parameters) boundaries within which the desired system functions are maintained.

Currently, managers of many arid and semi-arid ecosystems consider fire-promoting or fire-responsive alien species to be among their key management challenges. Most examples of changes in fire regime driven by alien plants have involved grasses, which appear to have dramatically increased the frequency[83,90] and intensity of fire.[86,91] There was also a recent study of shifts in the timing of fire as driven by the prevalence of alien fuels.[92] Few quantified examples exist of invasive species reducing the frequency or intensity of fire (but see Ref. 93) beyond a threshold such that ecosystem transformation occurs, although there are suggestions that many species have the potential to do so.[94]

Most studies of the relationship of nonnative species to fire do not evaluate thresholds of composition or process beyond which ecosystems would move out of a defined BRV or off of a trajectory that would keep them in a desired basin of attraction. Indeed, for most case studies evaluat-

ing invasive species and fire, the historical range of fire regime parameters of the system before invasion is unknown, but anecdotal evidence suggests that it is substantially outside the realm of the new fire regime that is being promoted by the invaders. For example, the annual Mediterranean grass *Bromus tectorum* (cheatgrass) has invaded the North American intermountain west where it is promoting frequent fire[83] to the detriment of native woody species cover.[95,96] While it has been demonstrated that individual fire events locally eliminate native woody species in this ecosystem (reviewed in Ref. 96), and recurrent fire at less than a five-year return interval further deplete native species,[83] neither the fire frequency required to maintain the cheatgrass-dominated state over decades nor the HRV of the desired condition are yet known. For cheatgrass in sagebrush ecosystems of the Great Basin, the desired condition is assumed to be one in which occasional (> 35-year return interval, see Ref. 96) fire could maintain a landscape balance between a more shrub-dominated (sagebrush) and more native perennial grass-dominated condition[97] with low cheatgrass abundance. Indeed, Baker[96] warns that fire return intervals for sagebrush-dominated ecosystems are likely very long, and thus cautions against using fire to try to keep senescent sagebrush systems from crossing thresholds of susceptibility to cheatgrass–fire transformation. A region-wide study is underway to evaluate thresholds of sagebrush, native perennial grass, and cheatgrass abundance within which fire can be used as a tool to maintain the system within a range of desired states (http://www.sagestep.org). This study is unique in its focus on what could be considered a TPC, although they do not use this terminology.

Since the impact of a species is a function of both its abundance and its unique per capita traits,[98] the latter in this case being fuel characteristics, it is possible to define on-the-ground thresholds of invader abundance that greatly increase the probability of the ecosystem moving to a new basin of attraction or to a potentially irreversible undesired condition. Likewise, it might also be possible to define environmental or disturbance regime BRV thresholds that limit invasion of the system by undesirable species. Such thresholds have been identified in a southeastern United States prairie ecosystem, where it was found that fire intervals of four years or less prohibited the establishment of an invasive shrub and intervals above this allowed the invader to become abundant and resistant to fire mortality.[93] Careful research could therefore guide identification of other BRV thresholds for ecosystems invaded or being threatened by fire-associated exotic species.

Thresholds for fire-transformed areas at the landscape scale also deserve greater study. For example, the negative effects of sagebrush ecosystem fragmentation via cheatgrass and fire have been documented[89] for breeding passerine birds (also see Ref. 99). These studies suggest thresholds of landscape scale transformation beyond which certain species disappear from the habitat. Our framework of BRV would therefore provide a basis for managers to assess the extent to which maintaining or restoring patches of desired vegetation within a regional landscape is essential to species persistence within that region.

BRV: fire and carbon

Carbon sequestration in natural systems has gained considerable attention due to its mitigating effect on climate change. With this attention has come increasing pressure on land managers to account for the effects of management actions on carbon stocks and emissions. The relatively new emphasis on carbon is above and beyond typical ecosystem management concerns (e.g., habitat and water quality), which may be strongly affected by fire. This focus on carbon may also be seen as more of a socio-political priority, rather than one explicitly relating to an ecological tipping point, but it lends itself to BRV concepts in several ways.

The frequency of fire as a natural process ranges from frequent, in systems such as grasslands and dry forests, to infrequent, in systems such as temperate wet forests. Infrequent fire systems are characterized by a fire event resetting the successional stage of the system, often called crown fire or stand renewing if the dominant overstory canopy is consumed. From a carbon perspective, this results in a direct flux of carbon to the atmosphere during the combustion process and an indirect flux from the decomposition of fire-killed vegetation. The immediate effects are often characterized as the system transitioning from carbon sink to source. However, over a period of time encompassing the broader HRV, vegetation recovers and the system transitions back to a sink.[100] For frequent fire systems, mortality rates are lower, typically resulting in reduced indirect carbon emissions compared to infrequent fire systems. Reduced mortality rates also allow the system to sequester the carbon emitted during the fire in a relatively short period of time.[101]

Because forests burn at varying intervals, the carbon stock and source-sink dynamics oscillate as a function of the frequency of disturbance, the primary difference being the amplitude of the oscillations.[102–105] Theoretically, the amount of carbon in a given forest could be maximized by eliminating disturbances, but this is unlikely to be indefinitely sustainable.[106] Fire exclusion in many forests with historically frequent fires has altered the vegetation structure in these ecosystems, resulting in a range of effects on carbon stocks.[105, 107–110] Regardless of the carbon-stock implications of fire-exclusion, the change in structure and associated fuel buildup has often pushed fire regimes from frequent, low-severity toward less frequent, high-severity fire, approaching what some would consider a TPC. Others might see this as typical of a naturally mixed-severity fire regime, although the extent to which this occurred prehistorically has yet to be determined.

Systematically restoring fire to many forest systems is typically achieved by targeting the mean structure associated with HRV through mechanical thinning and/or prescribed burning. These actions result in carbon removal (thinning) and emissions (burning, equipment, etc.), with the size of the carbon-stock reduction and emissions varying as a function of a number of factors, including treatment intensity.[104, 108, 111, 112] It has been argued that these treatments can yield a net carbon benefit because of the reduction in emissions and mortality associated with fire occurrence.[103, 113] However, some have argued that at the landscape scale, these treatments generally result in a net loss of carbon when compared to fire, because of the need to treat more land area than is potentially burned by wildfire.[104]

The concept of BRV thresholds in fire management can be applied to carbon storage in any fire-prone ecosystem, where the carbon stock in the system fluctuates across time and space as a function of the frequency and intensity of fire. In this context, a useful concept for framing the discussion is the carbon carrying capacity (CCC). The CCC of a system is the amount of carbon that can be stored in different age classes of vegetation under prevailing climatic and natural disturbance conditions, but excluding anthropogenic disturbances.[114, 115] This concept can serve as a benchmark against which the current carbon stock in a given location can be evaluated. Levels of carbon storage below CCC would reflect capacity lost due to past land use and management activities. Above the CCC would be a carbon saturation level prone to carbon emissions from the system given the occurrence of natural disturbances.[116]

In the case of fire and human intervention, a given ecosystem may be experiencing fires at a frequency higher or lower than the one under which it evolved. When fire is too frequent, serotinous, obligate seeding, and resprouting species can be negatively affected;[117] alternatively, when fire is too rare, community structure can be fundamentally altered resulting in a buildup of fuel and a fundamental shift in fire type.[118] Both ends of this spectrum could constitute boundaries for a particular ecosystem in a BRV framework. The implications of altered fire frequency range from substantial reductions in the carbon stock (e.g., grassland replacement of shrubland) with shortened fire intervals to substantial increases with lengthened fire intervals.[109, 110, 119] In cases where fire management for conservation seeks to reduce fire frequency to aid in ecosystem recovery, the potential exists for increasing the carbon stock closer to some natural carrying capacity. BRV thresholds and management triggers could therefore be identified based on ecological targets, carbon sequestration goals, and tradeoffs therein. For example, in systems where the goal is to restore fire as a more frequent process and the system is currently above the carbon carrying capacity, reductions in the carbon stock will be required.

The effectiveness of viewing the CCC of a natural system in the context of BRV is that it accounts for changes stemming from both climate and disturbance, shifts that may result in the range shifting upward or downward, expanding or contracting. An example where the range is likely to move up can be found in northeastern U.S. forests, where fire exclusion has caused a shift from open- to closed-canopy forest, resulting in mesofication of the system,[120] and climate change is projected to enhance future productivity.[121] The role of fire has diminished, and with predicted productivity increases, the upper range of carbon variation could potentially increase. Attempting to restore fire in these less flammable systems, simply because it has

been historically important (using HRV as the management target), could push carbon below a BRV threshold, as defined by prevailing climatic and natural disturbance conditions. In this case, crossing the BRV threshold for carbon (i.e., below the CCC of the system) may result in a shift in species composition (from more mesic to more xeric species), and it needs to be considered whether this also crosses a BRV threshold for ecosystem composition.

While, conceptually, the CCC can provide benchmarks for an appropriately sized carbon stock in a given system, it is imperative that we do not fall into the trap of having a single (or maximum) target carbon condition for every patch within the entire landscape, because it would neglect important ecosystem heterogeneity. The roles of disturbance variability and changing climate must be considered in defining an ecologically appropriate BRV for the carbon stock. As climate changes, system-level productivity is likely to be affected,[122] altering the amount of carbon that can be stored in the system, and fire frequency may change as temperature and precipitation change.[123–126] Accounting for multiple BRV thresholds can thus accommodate the role of both ecological disturbances and climate. Furthermore, understanding that desired carbon stocks will fluctuate and sometimes even conflict with conservation goals allows for tradeoffs in socio-ecological constraints, rather than seeking maximization of this one ecosystem service.

Implementation of BRV

Because historical dynamics of many ecosystems are only known for relatively short periods, and anticipating tipping points of change is difficult, the BRV framework we propose here will have to evolve as it is put into practice. It should be possible to adjust decision points that trigger management action as we learn more about the range of factors negatively affecting systems today and how they are likely to shift in the future (e.g., choosing a more conservative percentile in the tails of the distribution). This is especially true for our growing knowledge of ecological threshold dynamics,[37] but it also applies to shifting public perceptions of what types of changes are unacceptable.

The invasive species and carbon examples above examine the potential for using a BRV framework, but without the details of putting it into practice. To develop and implement BRV in a real-world

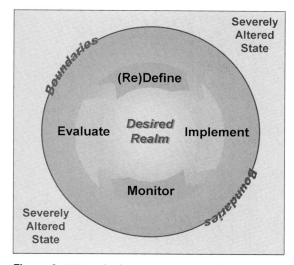

Figure 2. BRV and adaptive management. The traditional adaptive management feedback loop can be used to iteratively refine BRV by incorporating knowledge of desired heterogeneity in past landscape patterns and processes (HRV) and undesirable changes (TPC). Triggers for management activities would be identified at some distance in parameter space from the desired realm of dynamics, but prior to reaching boundaries.

setting, one would necessarily embed the framework in an adaptive feedback loop (Fig. 2). The traditional adaptive management cycle could thus emphasize ecosystem resilience and allow for entering into the BRV framework with limited knowledge through the following iterative stages: (1) *defining the context of the problem and desired future conditions*, including key drivers of change and potential future ranges of variation, measurable pattern/process targets and bounds for specific parameters, management triggers that precede bounds, feasible activities to achieve BRV goals, and experimental trials to confirm cause-effect relationships; (2) *implementing the plan* using passive landscape management to promote natural dynamics, active landscape management to facilitate desired heterogeneity, and active landscape management triggered to avoid severe alterations; (3) *monitoring outcomes of management and natural dynamics*, including actual management activities undertaken (versus planned), variation in key pattern/process metrics characterizing dynamics, and trends toward future management triggers; (4) *evaluating new information*, including causal links between management and landscape changes, observed natural stochasticity in dynamics, intended versus unin-

tended outcomes, and recent outside research on HRV and/or TPC in related systems; and (5) *beginning the process again* to refine and achieve BRV goals.

The up-front quantification of measurable pattern and process targets and bounds for specific parameters is admittedly challenging, and a review of the various methodologies used for each is beyond the scope of this paper. However, the previously cited HRV and TPC literature will provide useful guidance for this step, and a notable merger of approaches is described in Gillson and Duffin,[71] where a TPC for reductions in woody vegetation cover is evaluated in relation to fossil pollen records describing its HRV.

Where knowledge of ecosystem dynamics is lacking, simulation models are a valuable set of tools for predicting ecosystem states, trajectories, ranges of variation, and threshold transitions under various conditions.[127–129] In fact, simulation is one of the few approaches available for examining potential future scenarios and capturing rare occurrences of threshold dynamics. How tipping points in one ecosystem may relate to tipping points in another is also largely unknown. Simulation modeling thus allows for landscape-scale experiments to detect such thresholds in complex and stochastic systems.

While a fairly typical response to decision making under uncertainty is inaction and the call for more information, this is an increasingly futile option with regard to fire management and conservation. At a minimum, ranges of historical dynamics and potential thresholds may need to be hypothesized for BRV, based on knowledge of ecological tolerances for a few species, patterns, or processes (see Box 3). This provides a starting point for BRV, which can then be iteratively augmented with new information through time. In practice, the BRV framework should eventually result in a clear set of parameters for multiple ecosystem management goals, which would reflect the current ecosystem condition, distance from different thresholds, and management activities triggered to keep bounds from being crossed.

Concluding thoughts

Our goal has been to review current thinking on fire management and conservation, with an emphasis on maintaining inherent ecosystem resilience. In light of this goal, our contribution is oriented more

Box 3. Constructing BRV for chaparral shrublands

As a starting point for BRV, we use some of what is known about fire sizes and frequencies, as they relate to dominant chaparral shrubs and spawning habitat quality for a key anadromous fish.

Short fire-return intervals can restrict recruitment of plant species that recruit by seed, if they require a longer fire-free interval to reestablish a sufficient seed bank. Certain fire-dependent shrub species may thus be lost after frequent fires.[78] Environmental variation (e.g., precipitation gradients) can speed or slow the time to seed production, also affecting where a threshold in fire frequency exists. On the long interval extreme, these same species could conceivably be eliminated, as the seed bank dies off along with the parent plants in the absence of fire. The threshold on rapid fire frequencies is likely to be relatively sharp, whereas the long-interval extreme may be a more gradual system boundary. For chaparral shrublands of California, a short-interval threshold may exist in the ~ 15 year range; the long-interval extreme for chaparral, if it exists, is not known.

Small fires are not seen as a risk to many species. In contrast, large events can have severe effects on some ecosystems.[56] Great

For ecological thresholds at short fire intervals of ~ 15 years and large fire sizes ~ 68 km in width, one could identify boundaries for chaparral ecosystem dynamics, beyond which some management intervention should be considered. Overlain on HRV, such triggers would constitute outer edges of BRV. As more is learned about ecological responses (e.g., in parameter space), BRV can be adjusted accordingly.

distances to unburned seed sources may result from large fires, as well as other landscape-altering effects (e.g., debris flows). In chaparral shrublands, postfire seed dispersal is not limiting; however, effects of large fires on riparian areas can be substantial. Large fires and subsequent fine sediments can threaten endangered southern Steelhead, an anadromous fish species that spawns in creeks of chaparral-dominated landscapes. In key Steelhead conservation areas, management studies have thus used the 1000-year event (fires >68 km in width) as being events of concern.[79] The figure demonstrates the identification of these parameter space thresholds in HRV, along with the management triggers that precede them, constituting a minimal BRV framework for this system.

---

toward conserving ecosystems in some desired and historically relevant condition, rather than restoring badly degraded habitats or guiding new assemblages of species into novel ecosystems. To build on existing theory and management approaches, we have suggested the merger of two decision-making frameworks from different fire-prone parts of the world.

Although climate change will have diverse effects and sometimes result in novel distributions of fire,[130] future fire activity is likely to have substantial overlap with historical patterns, and many topographic microclimates should be maintained

at fine scales.[131] HRV in fire therefore provides an ongoing and useful context for decision making. It is, however, the extremes that often matter most,[132] and this requires an emphasis on thresholds. Incorporating known or expected tipping points in key disturbance regime parameters is therefore an important contribution from the TPC framework, as is the inclusion of socially unacceptable levels of change. Social and ecological goals will, of course, be at odds sometimes, so these too may require trade-offs and value judgments.

The uncertainty of the future climate space for a given geographic location may at first make the facilitation of system-level resilience seem intractable. However, the BRV concept, used as part of the decision matrix for managing under uncertainty, can provide assistance in allocating conservation resources. For multiple conservation areas, one could first assess system state within BRV and the associated risk of crossing some critical resilience threshold. The conservation priority of different areas, combined with knowledge of their BRV status, can then serve as a means for allocating scarce conservation resources to priority areas with the greatest capacity for resilience. Given the scarcity of conservation dollars and projected regional changes, the goal of restoring some ecosystems throughout the space they currently occupy may be unattainable. However, areas that have a high conservation priority (e.g., endemic populations present) and have yet to approach BRV thresholds (burned by uncharacteristically severe wildfire) could be selected as a target for resource investment to mitigate the risk that some conservation value will be lost due to a fire-driven change.

In some cases, the ecosystem management goal may be to appropriately target a preferred alternative state or basin of attraction outside BRV. In the context of fire management, a conservative approach might involve targeted fire-exclusion in a particular burned area to ensure an opportunity for colonization by new species likely to disperse into the area and survive under new conditions. Allowing natural colonization is likely to be most effective in areas where the candidate group of species is in relatively close proximity, such as along a steep elevation gradient. In areas where a desirable set of new candidate species is not in close proximity, this may require human intervention.[15,17] In a changing climate, however, which human intervention should be taken will depend on knowledge of a preferred alternative state (and the nearby basins of attraction). Therefore, decisions should link directly back to the historical range of ecosystem dynamics and the thresholds therein (i.e., BRV).

Although we have proposed the application of BRV in an adaptive management context above, BRV is currently more of a conceptual framework than one ready for widespread application. We have not directly addressed the issue of scale, and this will have to be explored in depth if BRV is to take hold in decision making. The patterns and processes in question would imply use over relatively broad landscape extents. Very localized and high-value conservation targets, such as iconic stands of certain trees or unique endemic habitats, should simply be preserved if possible. Over broader scales, landscape heterogeneity must provide some linkage to ecosystem resilience, both past and future. This heterogeneity is structured in space and time, however, and its most careful application will also incorporate ecological thresholds and the management triggers that should precede them.

Acknowledgments

A portion of this work was conducted by M.A.M. while a Center Fellow at the National Center for Ecological Analysis and Synthesis, a center funded by NSF (Grant #EF-0553768), the University of California, Santa Barbara, and the State of California. We thank E. Batllori for help with processing data for fire probability distributions.

Conflicts of interest

The authors declare no conflicts of interest.

References

1. Cissel, J.H., F.J. Swanson, W.A. McKee & A.L. Burditt. 1994. Using the past to plan the future in the Pacific Northwest. *J. For.* **92:** 30–46.
2. Kaufmann, M.R., R.T. Graham, D.A.J. Boyce, *et al.* 1994. *An Ecological Basis for Ecosystem Management.* Gen. Tech. Rep. RM 246. Fort Collins, CO: USDA Forest Service, Rocky Mountain Forest and Range Experiment Station.
3. Swanson, F.J., J.A. Jones, D.O. Wallin & J.H. Cissel. 1994. "Natural variability: implications for ecosystem management." In *Volume II: Ecosystem Management: Principles and Applications.* M.E. Jensen & P.S. Bourgeron, Eds: 80–94. Portland, OR: USDA Forest Service, Pacific Northwest Research Station.
4. Aplet, G. & W.S. Keeton. 1999. "Application of historical range of variability concepts to biodiversity conservation." In *Practical Approaches to the Conservation of Biological Diversity.* R.K. Baydack, H. Campa & J.B. Haufler, Eds.: 71–86. New York, NY: Island Press.
5. Landres, P.B., P. Morgan & F.J. Swanson. 1999. Overview and use of natural variability concepts in managing ecological systems. *Ecol. Appl.* **9:** 1179–1188.
6. Millar, C.I. & W.B. Woolfenden. 1999. The role of climate change in interpreting historical variability. *Ecol. Appl.* **9:** 1207–1216.
7. Swetnam, T.W., C.D. Allen & J.L. Betancourt. 1999. Applied historical ecology: using the past to manage for the future. *Ecol. Appl.* **9:** 1189–1206.
8. Egan, D. & E.A. Howell Eds., 2001. *The Historical Ecology Handbook: A Restorationist's Guide to Reference Ecosystems.* Washington, DC: Island Press.

9. Keane, R.E., R.A. Parsons & P.F. Hessburg. 2002. Estimating historical range and variation of landscape patch dynamics: limitations of the simulation approach. *Ecol. Model.* **151:** 29–49.

10. Veblen, T.T. 2003. Historic range of variability of mountain forest ecosystems: concepts and applications. *For. Chron.* **79:** 223–226.

11. Keane, R.E., P.F. Hessburg, P.B. Landres & F.J. Swanson. 2009. The use of historical range and variability (HRV) in landscape management. *For. Ecol. Manag.* **258:** 1025–1037.

12. Romme, W.H.J., J.A. Wiens, & H.D. Safford. 2012. "Setting the stage: theoretical and conceptual background of historical range of variation." In *Historical Environmental Variation in Conservation and Natural Resource Management.* J.A. Wiens, G.D. Hayward, H.D. Safford & C.M. Giffen, Eds.: 3–18. Chichester: John Wiley and Sons, Ltd.

13. Holling, C.S. & G.K. Meffe. 1996. Command and control and the pathology of natural resource management. *Conserv. Biol.* **10:** 328–337.

14. Klausmeyer, K.R., M.R. Shaw, J.B. MacKenzie & D.R. Cameron. 2011. Landscape-scale indicators of biodiversity's vulnerability to climate change. *Ecosphere* **2:** art88.

15. Millar, C.I., N.L. Stephenson & S.L. Stephens. 2007. Climate change and forests of the future: managing in the face of uncertainty. *Ecol. Appl.* **17:** 2145–2151.

16. North, M., P. Stine, K. O'Hara, W. Zielinski & S. Stephens. 2009. *An Ecosystem Management Strategy for Sierran Mixed-Conifer Forests.* Gen. Tech. Rep. PSW-GTR-220. Albany, CA: USDA Forest Service, Pacific Southwest Research Station.

17. Stephens, S.L., C.I. Millar & B.M. Collins. 2010. Operational approaches to managing forests of the future in Mediterranean regions within a context of changing climates. *Environ. Res. Lett.* **5:** 024003.

18. Rogers, K. & H.C. Biggs. 1999. Integrating indicators, endpoints and value systems in strategic management of the Kruger National Park. *Freshwater Biol.* **41:** 439–451.

19. Johnstone, J.F., F.S. Chapin, T.N. Hollingsworth, *et al.* 2010. Fire, climate change, and forest resilience in interior Alaska. *Canad. J. For. Res.* **40:** 1302–1312.

20. Kitzberger, T., E. Aráoz, J.H. Gowda, *et al.* 2012. Decreases in fire spread probability with forest age promotes alternative community states, reduced resilience to climate variability and large fire regime shifts. *Ecosystems* **15:** 97–112.

21. Moritz, M.A., P.F. Hessburg & N.A. Povak. 2011. "Native fire regimes and landscape resilience." In *The Landscape Ecology of Fire.* Ecological Studies 23. D. McKenzie, C. Miller & D.A. Falk, Eds.: 51–86. Netherlands: Springer.

22. Whitlock, C., P.E. Higuera, D.B. McWethy & C.E. Briles. 2010. Paleoecological perspectives on fire ecology: revisiting the fire-regime concept. *Open Ecol. J.* **3:** 6–23.

23. De Leo, G.A. & S. Levin. 1997. The multifaceted aspects of ecosystem integrity. *Conserv. Ecol.* **1:**3. http://www.consecol.org/vol1/iss1/art3/

24. Pratt, S.D., L. Holsinger & R.E. Keane. 2006. "Modeling historical reference conditions for vegetation and fire regimes using simulation modeling." In *The LANDFIRE Prototype Project: Nationally Consistent and Locally Relevant Geospatial Data for Wildland Fire Management.* M.G.

Rollins & C. Frame, Eds: 277–314. Fort Collins, CO: USDA Forest Service, Rocky Mountain Research Station.

25. Keane, R.E., L.M. Holsinger, R.A. Parsons & K. Gray. 2008. Climate change effects on historical range and variability of two large landscapes in western Montana, USA. *For. Ecol. Manag.* **254:** 375–389.

26. Bormann, F.H. & G.E. Likens. 1979. *Pattern and Process in a Forested Ecosystem.* New York: Springer.

27. Wu, J. & O.L. Loucks. 1995. From balance of nature to hierarchical patch dynamics: a paradigm shift in ecology. *Qtly. Rev. Biol.* **70:** 439–466.

28. McGarigal, K. & W.H. Romme. 2012. "Modeling historical range of variability at a range of scales: an example application." In *Historical Environmental Variation in Conservation and Natural Resource Management.* J.A. Wiens, G.D. Hayward, H.D. Safford & C.M. Giffen, Eds.: 128–146. Chichester: John Wiley and Sons, Ltd.

29. Holling, C.S. 1973. Resilience and stability of ecological systems. *Annu. Rev. Ecol. Syst.* **4:** 1–23.

30. Folke, C., S. Carpenter, B. Walker, *et al.* 2004. Regime shifts, resilience, and biodiversity in ecosystem management. *Annu. Rev. Ecol. Evol. Syst.* **35:** 557–581.

31. Walker, B., A. Kinzig & J. Langridge. 1999. Plant attribute diversity, resilience, and ecosystem function: the nature and significance of dominant and minor species. *Ecosystems* **2:** 95–113.

32. Scheffer, M. & S.R. Carpenter. 2003. Catastrophic regime shifts in ecosystems: linking theory to observation. *Trends Ecol. Evol.* **18:** 648–656.

33. Anderies, J.M., M.A. Janssen & B.H. Walker. 2002. Grazing management, resilience, and the dynamics of a fire-driven rangeland system. *Ecosystems* **5:** 23–44.

34. Myers, R.L. 1985. Fire and the dynamic relationship between Florida sandhill and sand pine scrub vegetation. *Bull. Torrey Bot. Club* **112:** 241–252.

35. Odion, D.C., M.A. Moritz & D.A. DellaSala. 2010. Alternative community states maintained by fire in the Klamath Mountains, USA. *J. Ecol.* **98:** 96–105.

36. Carpenter, S., B. Walker, J.M. Anderies & N. Abel. 2001. From metaphor to measurement: resilience of what to what? *Ecosystems* **4:** 765–781.

37. Suding, K.N. & R.J. Hobbs. 2009. Threshold models in restoration and conservation: a developing framework. *Trends Ecol. Evol.* **24:** 271–279.

38. Hughes, T.P., N.A.J. Graham, J.B.C. Jackson, *et al.* 2010. Rising to the challenge of sustaining coral reef resilience. *Trends Ecol. Evol.* **25:** 633–642.

39. D'Antonio, C.M. & P.M. Vitousek. 1992. Biological invasions by exotic grasses, the grass/fire cycle, and global change. *Annu. Rev. Ecol. Syst.* **23:** 63–87.

40. van der Heide, T., E.H. van Nes, G.W. Geerling *et al.* 2007. Positive feedbacks in seagrass ecosystems: implications for success in conservation and restoration. *Ecosystems* **10:** 1311–1322.

41. Carpenter, S.R., D. Ludwig & W.A. Brock. 1999. Management of eutrophication for lakes subject to potentially irreversible change. *Ecol. Appl.* **9:** 751–771.

42. White, P.S. & J.L. Walker. 1997. Approximating nature's variation: selecting and using reference information in restoration ecology. *Restor. Ecol.* **5:** 338–349.

43. Connell, J.H. & W.P. Sousa. 1983. On the evidence needed to judge ecological stability or persistence. *Am. Nat.* **121:** 789–824.

44. Nystrom, M. & C. Folke. 2001. Spatial resilience of coral reefs. *Ecosystems* **4:** 406–417.

45. Peters, D.P.C., B.T. Bestelmeyer, J.E. Herrick, *et al.* 2006. Disentangling complex landscapes: new insights into arid and semiarid system dynamics. *BioScience* **56:** 491–501.

46. Chapin, F.S. & A.M. Starfield. 1997. Time lags and novel ecosystems in response to transient climatic change in arctic Alaska. *Clim. Change* **35:** 449–461.

47. Duffy, J.E. 2002. Biodiversity and ecosystem function: the consumer connection. *Oikos* **99:** 201–219.

48. Jackson, J.B.C. *et al.* 2001. Historical overfishing and the recent collapse of coastal ecosystems. *Science* **293:** 629–638.

49. McWethy, D.B. *et al.* 2010. Rapid landscape transformation in South Island, New Zealand, following initial Polynesian settlement. *Proc. Natl. Acad. Sci. USA* **107:** 21343–21348.

50. Chapin, F.S. *et al.* 2000. Consequences of changing biodiversity. *Nature* **405:** 234–242.

51. Norberg, J., D.P. Swaney, J. Dushoff, *et al.* 2001. Phenotypic diversity and ecosystem functioning in changing environments: a theoretical framework. *Proc. Natl. Acad. Sci. USA* **98:** 11376–11381.

52. Suding, K.N., I.W. Ashton, H. Bechtold, *et al.* 2008. Plant and microbe contribution to community resilience in a directionally changing environment. *Ecol. Monograp.* **78:** 313–329.

53. Elmqvist, T., C. Folke, M. Nystrom, *et al.* 2003. Response diversity, ecosystem change, and resilience. *Front. Ecol. Environ.* **1:** 488–494.

54. Hilborn, R., T.P. Quinn, D.E. Schindler & D.E. Rogers. 2003. Biocomplexity and fisheries sustainability. *Proc. Natl. Acad. Sci. USA* **100:** 6564–6568.

55. Flint, L.E. & A.L. Flint. 2012. Downscaling future climate scenarios to fine scales for hydrologic and ecological modeling and analysis. *Ecol. Proces.* **1:** 1–15.

56. Romme, W.H., E.H. Everham, L.E. Frelich, *et al.* 1998. Are large infrequent disturbances qualitatively different from small frequent disturbances? *Ecosystems* **1:** 524–534.

57. Schmidt, K.M., J.P. Menakis, C.C. Hardy, *et al.* 2002. *Development of Coarse-Scale Spatial Data for Wildland Fire and Fuel Management.* Gen. Tech. Rep. RMRS-GTR-87. Fort Collins, CO: USDA Forest Service, Rocky Mountain Research Station.

58. Hessburg, P.F., B.G. Smith & R.B. Salter. 1999. Detecting change in forest spatial patterns from reference conditions. *Ecol. Appl.* **9:** 1232–1252.

59. Gärtner, S., K.M. Reynolds, P.F. Hessburg, *et al.* 2008. Decision support for evaluating landscape departure and prioritizing forest management activities in a changing environment. *For. Ecol. Manag.* **256:** 1666–1676.

60. Williams, J.W. & S.T. Jackson. 2007. Novel climates, no-analog communities, and ecological surprises. *Front. Ecol. Environ.* **5:** 475–482.

61. Seastedt, T.R., R.J. Hobbs & K.N. Suding. 2008. Management of novel ecosystems: are novel approaches required? *Front. Ecol. Environ.* **6:** 547–553.

62. Hiers, J.K., R.J. Mitchell, A. Barnett, *et al.* 2012. The dynamic reference concept: measuring restoration success in a rapidly changing no-analogue future. *Ecol. Restor.* **30:** 27–36.

63. Hessburg, P.F., K.M. Reynolds, R.B. Salter, *et al.* 2013. Landscape evaluation for restoration planning on the Okanogan-Wenatchee National Forest, USA. *Sustainability* **5:** 805–840.

64. Hannah, L., L. Flint, A. Syphard, *et al.* Fine-scale modeling of vegetation response to climate change. *Trends Ecol. Evol.* In review.

65. Krawchuk, M.A. & M.A. Moritz. 2012. *Fire and Climate Change in California.* California Energy Commission. Publication number: CEC-500-2012-026.

66. Parrish, J.D., D.P. Braun & R.S. Unnasch. 2003. Are we conserving what we say we are? Measuring ecological integrity within protected areas. *BioScience* **53:** 851–860.

67. Groffman, P.M. *et al.* 2006. Ecological thresholds: the key to successful environmental management or an important concept with no practical application? *Ecosystems* **9:** 1–13.

68. Martin, J., M.C. Runge, J.D. Nichols, *et al.* 2009. Structured decision making as a conceptual framework to identify thresholds for conservation and management. *Ecol. Appl.* **19:** 1079–1090.

69. Samhouri, J.F., P.S. Levin & C.H. Ainsworth. 2010. Identifying thresholds for ecosystem-based management. *PLoS ONE* **5:** e8907.

70. Rogers, K.H. 2003. "Adopting a heterogeneity paradigm: implications for management of protected savannas." In *The Kruger Experience: Ecology and Management of Savanna Heterogeneity.* du Toit, J.T., K.H. Rogers & H.C. Biggs, Eds.: 41–58. Washington, DC: Island Press.

71. Gillson, L. & K.I Duffin. 2007. Thresholds of potential concern as benchmarks in the management of African savannahs. *Philos. Trans. R. Soc. Lond. B Biol. Sci.* **362:** 309–319.

72. Biggs, H.C. & K.M. Rogers. 2003. "An adaptive system to link science, monitoring and management in practice." In *The Kruger Experience: Ecology and Management of Savanna Heterogeneity.* du Toit, J.T., K.H. Rogers & H.C. Biggs, Eds.: 59–80. Washington, DC: Island Press.

73. van Wilgen, B.W., N. Govender & H.C. Biggs. 2007. The contribution of fire research to fire management: a critical review of a long-term experiment in the Kruger National Park, South Africa. *Int. J. Wildland Fire* **16:** 519–530.

74. Roux, D. & L.C. Foxcroft. 2011. The development and application of Strategic Adaptive Management (SAM) within South African National Parks. *Koedoe* **53:**1049.

75. Van Wilgen, B.W., N. Govender, G.G. Forsyth & T. Kraaij. 2011. Towards adaptive fire management for biodiversity conservation: experience in South African National Parks. *Koedoe* **53:**982.

76. Chapple, R.S., D. Ramp, R.A. Bradstock, *et al.* 2011. Integrating science into management of ecosystems in the Greater Blue Mountains. *Environ. Manag.* **48:** 659–674.

77. Biggs, H., S. Ferreira, S. Freitag-Ronaldson & R. Grant-Biggs. 2011. Taking stock after a decade: does the 'thresholds of potential concern' concept need a socioecological revamp? *Koedoe* **53:** 1002.

78. Zedler, P.H., C.R. Gautier & G.S. McMaster. 1983. Vegetation change in response to extreme events: the effect of a short interval between fires in California chaparral and coastal scrub. *Ecology* **64:** 809–818.

79. Boughton, D.A. *et al.* 2007. *Viability Criteria for Steelhead of the South-central and Southern California Coast.* NOAA Tech. Mem. Series Draft NOAA-TM-NMFS-SWFSC-407. Santa Cruz, CA: NMFS.

80. Richardson D.M., N. Allsopp, C.M. D'Antonio, *et al.* 2000. Plant invasions: the role of mutualisms. *Biol. Rev.* **75:** 65–93.

81. Rossitor, N.A., S.A. Setterfield, M.M. Douglas & L.B. Hutley. 2003. Testing the grass-fire cycle: alien grass invasion in the tropical savannas of northern Australia. *Divers. Distrib.* **9:** 169–176.

82. Brooks, M.L., C.M. D'Antonio, D.M. Richardson, *et al.* 2004. Effects of invasive alien plants on fire regimes. *BioScience* **54:** 677–688.

83. Whisenant S.G. 1990. *Changing Fire Frequencies on Idaho's Snake River Plains: Ecological and Management Implications. Gen. Tech. Rep.* INT-276. Logan, UT: USDA Forest Service, Intermountain Research Center.

84. Hughes R.F., P.M. Vitousek & J.T. Tunison. 1991. Exotic grass invasion and fire in the seasonal submontane zone of Hawaii. *Ecology* **72:** 743–746.

85. D'Antonio, C.M., R.F. Hughes & J.T. Tunison. 2011. Long-term impacts of invasive grasses and subsequent fire in seasonally dry Hawaiian woodlands. *Ecol. Appl.* **21:** 1617–1628.

86. Setterfield, S.A., N.A. Rossitor-Ranchor, L.B. Hutley, *et al.* 2010. Turning up the heat: the impacts of *Andropogon gayanus* (gamba grass) invasion on fire behaviour in northern Australia savannas. *Divers. Distrib.* **16:** 854–861.

87. Obrist, D., E.H. Delucia & J.A. ARnone III. 2003. Consequences of wildfire on ecosystem CO2 and water vapour fluxes in the Great Basin. *Glob. Change Biol.* **9:** 563–574.

88. Bradley, B., R.A. Houghton, J.F. Mustard & S.P. Hamburg. 2006. Invasive grass reduces aboveground carbon stocks in shrublands of the western US. *Glob. Change Biol.* **12:** 1815–1822.

89. Knick, S. & J. Rotenberry. 1995. Landscape characteristics of fragmented shrubsteppe habitats and breeding passerine birds. *Conserv. Biol.* **9:** 1059–1071.

90. Tunsion, J.T., R. Loh & C.M. D'Antonio. 2001. "Fire, grass invasions and revegetation of burned areas in Hawaii Volcanoes National Park." In *Proceedings of the Invasive Species Workshop: The Role of Fire in the Controls and Spread of Invasive Species.* K.E. Galley & T.P. Wilson, Eds.: 122–131. Tall Timbers Research Station Publication No. 11, Lawrence Kansas: Allen Press.

91. Lippincott, C.L. 2000. Effects of *Imperata cylindrica* (L.) Beauv. (cogongrass) invasion on fire regime in Florida Sandhill (USA). *Nat. Areas J.* **20:** 140–149.

92. Balch, J.K., B.A. Bradley, C.M. D'Antonio & J. Gómez-Dans. 2013. Introduced annual grass increases regional fire activity across the arid western USA (1980–2009). *Glob. Change Biol.* **19:** 173–183.

93. Stevens, J.T. & B. Beckage. 2009. Fire feedbacks facilitate invasion of pine savannas by Brazilian pepper (*Schinus terebinthifolius*). *New Phytol.* **184:** 365–375.

94. D'Antonio C.M. 2000. "Fire, plant invasions, and global changes." In *Invasive Species in a Changing World.* H.A. Mooney & R.J. Hobbs, Eds.: 65–93. Washington, DC: Island Press.

95. Knick, S. 1999. Requiem for an ecosystem. *Northwest Sci.* **73:** 53–57.

96. Baker, W.L. 2006. Fire and restoration of sagebrush ecosystems. *Wildlife Soc. Bull.* **34:** 177–185.

97. D'Antonio, C.M., J.C. Chambers, R. Loh & J.T. Tunison. 2009. "Applying ecological concepts to the management of widespread grass invasions." In *Management of Invasive Weeds* Inderjit, Ed.: 123–149. New York: Springer.

98. Parker, I., D. Simberloff, W.M. Lonsdale, *et al.* 1999. Impact: toward a framework for understanding the ecological effects of invaders. *Biol. Invas.* **1:** 3–19.

99. Knick, S.T., D.S. Dobkin, J.T. Rotenberry, *et al.* 2003. Teetering on the edge or too late? Conservation and research issues for avifauna of sagebrush habitats. *Condor* **105:** 611–634.

100. Kashian, D.M., W.H. Romme, D.B. Tinker, *et al.* 2006. Carbon storage on landscapes with stand-replacing fires. *BioScience* **56:** 598–606.

101. Hurteau, M.D. & M. North. 2010. Carbon recovery rates following different wildfire risk mitigation treatments. *For. Ecol. Manag.* **260:** 930–937.

102. Beringer, J., L.B. Hutley, N.J. Tapper & L.A. Cernusak. 2007. Savanna fires and their impact on net ecosystem productivity in North Australia. *Glob. Change Biol.* **13:** 990–1004.

103. Hurteau, M. & M. North. 2009. Fuel treatment effects on tree-based forest carbon storage and emissions under modeled wildfire scenarios. *Front. Ecol. Environ.* **7:** 409–414.

104. Mitchell, S.R., M.E. Harmon & K.E.B. O'Connell. 2009. Forest fuel reduction alters fire severity and long-term carbon storage in three Pacific Northwest ecosystems. *Ecol. Appl.* **19:** 643–655.

105. Hurteau, M.D. & M.L. Brooks. 2011. Short- and long-term effects of fire on carbon in US dry temperate forest systems. *BioScience* **61:** 139–146.

106. Hudiburg, T., B. Law, D.P. Turner, *et al.* 2009. Carbon dynamics of Oregon and Northern California forests and potential land-based carbon storage. *Ecol. Appl.* **19:** 163–180.

107. Fellows, A.W. & M.L. Goulden. 2008. Has fire suppression increased the amount of carbon stored in western U.S. forests? *Geophys. Res. Lett.* **35:** L12404.

108. North, M., M. Hurteau & J. Innes. 2009. Fire suppression and fuels treatment effects on mixed-conifer carbon stocks and emissions. *Ecol. Appl.* **19:** 1385–1396.

109. Hurteau, M.D., M.T. Stoddard & P.Z. Fule. 2011. The carbon costs of mitigating high-severity wildfire in southwestern ponderosa pine. *Glob. Change Biol.* **17:** 1516–1521.

110. Collins, B.M., R.G. Everett & S.L. Stephens. 2011. Impacts of fire exclusion and recent managed fire on forest structure in old growth Sierra Nevada mixed-conifer forests. *Ecosphere* **2:** art51.

111. Finkral, A.J. & A.M. Evans. 2008. The effects of a thinning treatment on carbon stocks in a northern Arizona ponderosa pine forest. *For. Ecol. Manag.* **255:** 2743–2750.

112. Stephens, S.L., J.J. Moghaddas, B.R. Hartsough, *et al.* 2009. Fuel treatment effects on stand-level carbon pools, treatment-related emissions, and fire risk in a Sierra Nevada mixed-conifer forest. *Canad. J. For. Res.* **39:** 1538–1547.

113. Hurteau, M.D., G.W. Koch & B.A. Hungate. 2008. Carbon protection and fire risk reduction: toward a full accounting of forest carbon offsets. *Front. Ecol. Environ.* **6:** 493–498.

114. Gupta, R.K. & D.L.N. Rao. 1994. Potential of wastelands for sequestering carbon by reforestation. *Curr. Sci.* **66:** 378–380.

115. Keith, H., B.G. Mackey & D.B. Lindenmayer. 2009. Re-evaluation of forest biomass carbon stocks and lessons from the world's most carbon-dense forests. *Proc. Natl. Acad. Sci. USA* **106:** 11635–11640.

116. Moroni, M.T., T.H. Kelley & M.L. McLarin, 2010. Carbon in trees in Tasmanian State Forest. *Int. J. For. Res.* **25:**129–146.

117. Syphard, A.D., V.C. Radeloff, T.J. Hawbaker & S.I. Stewart. 2009. Conservation threats due to human-caused increases in fire frequency in Mediterranean-climate ecosystems. *Conserv. Biol.* **23:** 758–769.

118. Stephens, S.L. 1998. Evaluation of the effects of fuels treatments on potential fire behaviour in Sierra Nevada mixed-conifer forests. *For. Ecol. Manag.* **105:** 21–35.

119. Kaye, J.P., J. Romanya & V.R. Vallejo. 2010. Plant and soil carbon accumulation following fire in Mediterranean woodlands in Spain. *Oecologia* **164:** 533–543.

120. Nowacki, G.J. & M.D. Abrams. 2008. The demise of fire and "mesophication" of forest in the eastern Unted States. *BioScience* **58:** 123–138.

121. Campbell, J., D. Donato, D. Azuma & B. Law. 2007. Pyrogenic carbon emission from a large wildfire in Oregon, United States. *J. Geophys. Res.* **112:** G04014.

122. Beer, C. *et al.* 2010. Terrestrial gross carbon dioxide uptake: global distribution and covariation with climate. *Science* **329:** 834–838.

123. Westerling, A.L. & B.P. Bryant. 2008. Climate change and wildfire in California. *Clim. Change* **87**(Suppl. 1): S231–S249.

124. Krawchuk, M.A. & M.A. Moritz. 2011. Constraints on global fire activity vary across a resource gradient. *Ecology* **92:** 121–132.

125. Westerling, A.L., M.G. Turner, E.A.H. Smithwick, *et al.* 2011. Continued warming could transform greater Yellowstone fire regimes by mid-21st century. *Proc. Natl. Acad. Sci. USA* **108:** 13165–13170.

126. Moritz, M.A., M.-A. Parisien, E. Batllori, *et al.* 2012. Climate change and disruptions to global fire activity. *Ecosphere* **3:** 49.

127. Keane, R.E. & E. Karau. 2010. Evaluating the ecological benefits of wildfire by integrating fire and ecosystem simulation models. *Ecol. Model.* **221:** 1162–1172.

128. Scholes, R.J. & J.M. Kruger. 2011. A framework for deriving and triggering thresholds for management intervention in uncertain, varying and time-lagged systems. *Koedoe* **53:** 987.

129. Keane, R.E. 2012. "Creating historical range of variation (HRV) time series using landscape modeling: overview and issues." In *Historical Environmental Variation in Conservation and Natural Resource Management.* J.A. Wiens, G.D. Hayward, H.D. Safford & C.M. Giffen, Eds.: 113–127. Chichester: John Wiley and Sons, Ltd.

130. Krawchuk, M.A., M.A. Moritz, M.-A. Parisien, *et al.* 2009. Global pyrogeography: the current and future distribution of wildfire. *PLoS ONE* **4:** e5102.

131. Dobrowski, S.Z., 2011. A climatic basis for microrefugia: the influence of terrain on climate. *Glob. Change Biol.* **17:** 1022–1035.

132. Jentsch, A., J. Kreyling & C. Beierkuhnlein. 2007. A new generation of climate-change experiments: events, not trends. *Front. Ecol. Environ.* **5:** 365–374.